Dawn of Modern Science

Dawn of
Modern Science

FROM THE ARABS TO LEONARDO DA VINCI

Thomas Goldstein

HOUGHTON MIFFLIN COMPANY BOSTON
1980

Library of Congress Cataloging in Publication Data

Goldstein, Thomas.
    Dawn of modern science.
    Bibliography: p.
    Includes index.
    1. Science, Renaissance.   I. Title.
Q125.2.G64   509'.024   79-23753
ISBN 0-395-26298-4

Printed in the United States of America
V 10 9 8 7 6 5 4 3 2 1

Picture credits appear on page 297.

. . . Laudato si, mi Signore, cum tucte le tue creature,
spetialmente messor lo frate sole,
lo quale jorna, et illumini per lui;
et ellu è bellu e radiante cum grande splendore;
de Te, Altissimo, porta significatione. . .

Laudato si, mi Signore, per frate focu,
per lo quale ennallumini la nocte,
et ello è bellu, et jucundo, et robustoso et forte.
Laudato si, mi Signore, per sora nostra matre terra,
la quale ne sustenta e governa,
e produce diversi fructi, con coloriti fiori et herba. . .

Laudate et benedicete mi Signore, e rengratiate,
e serviteli cum grande humilitate.

. . . Be praised, my Lord, with all Thy creatures,
above all else lord brother sun,
who brings the day, through whom Thou givest light;
and he is beautiful and radiant with great splendor;
and he gives witness of Thee, Most High.

Be praised, my Lord, for brother fire,
through whom Thou lightest up the night,
and he is beautiful, and kind, and vigorous and strong.
Be praised, my Lord, for our sister mother earth,
who feeds and governs us,
and brings forth diverse fruit, and grass and many-colored flowers.

Let's praise and bless my Lord, and give Him thanks,
and serve Him in profound humility.

St. Francis, *The Canticle of the Sun*

Contents

Spinello Aretino's fresco, THE FOUNDING OF ALESSANDRIA BY POPE
ALEXANDER III (1407), in the city hall (Palazzo Pubblico) of Siena, exudes the
pride of the town in the achievements of its citizens, as well as in the use of
newfangled technological devices: Pope Alexander III (1159–81) was a scion of
the Sienese Bandinelli family; his founding of the North Italian fortress city of
Alessandria helped to protect Italy against the invasions of the German emperor
Frederick I. But the use of the pulley, driven by a crank, for hauling bricks to
the top of a city wall clearly reflects the vitality of Siena's own urban construction program, with its massive walls being built from the thirteenth to the
fifteenth century.

Introduction

Like many a reader, I approached the history of science as someone with an interest in the humanities and virtually no grounding in the natural sciences. My only link was my work in the scientific background of the discovery of America, involving the cosmological ideas of the later Middle Ages and, in particular, the geographic thought of the Renaissance. Yet these studies, though perfectly satisfying in themselves, had in no way changed my basic outlook on science — other than decidedly heightening my respect for the intellectual powers of a prescientific age.

Rather typically, I believe, mine was not merely a one-sided attitude by which I blocked out a vast field of knowledge as nonexistent for my intellectual interests. As I think back, I realize that I grew up with a profound distrust, not to say hostility, toward science and its modern twin, technology. To my essentially romantic soul science seemed the great destroyer, Siva, the unfeeling foe of scenic beauty and poetic simplicity, of the graceful life styles of a less purpose-oriented past and their architectural and decorative expressions.

By the same token, my attitude was characteristically ambivalent. I know now that the intrinsic ambivalence of the modern environmentalists is part of a centuries-old, highly respectable tradition of fears and protests, tempered by the recognition of science's undeniable human benefits. If our eyes have been sharpened by our twentieth-century experience for science's depersonalizing and destructive effects, we still acknowledge — and in any emergency welcome — the benefits of modern medical technology, or science's huge potential for social liberation, whether by freeing women from wasteful home drudgery or by offering developing nations a better access to economic prosperity and public health.

It was precisely this inner ambivalence, this Siva-Vishnu quality, that has always inspired a peculiar blend of awe, a mixture of apprehension and admiration, among those who witnessed science's dynamic rise, from the Medieval citizens terrified by the experiments of the alchemists to the early nineteenth-century Luddites and Romantics, down to the modern television audiences watching the mad scientist about to blow up the world. Throughout its rise to power glaring ambiguities have made it difficult for the contemporaries to recognize that science, despite our tendency to see it in simple moral terms, is neither "good" nor "bad" but an impersonal force as morally neutral as a computer or a machine, whose ethical value depends in essence on our uses. Perhaps behind these age-old fears there always lurked some presentiment about science's destructive potential, of which our age must face the full extent. I had to learn that our notorious ambivalence toward science is due to science's own ethical ambivalence — or, rather, its essential indifference toward its good or bad effects on humanity, its nature as an artifact, devoid of the human qualities we tend to project into it. If these were philosophically self-evident truths, I happened to learn them from the explicit evidence of history.

*　*　*

Yet to see science as the unique historical phenomenon it is does not mean robbing it of its attractions. While working on this book, I found that the historic approach gave me an access to scientific problems and their intrinsic fascination that had been closed to me before. With my humanistic bend of mind I had been led in by the wrong door, as it were, by teachers who had insisted on presenting science according to certain rigid traditional priorities, instead of appealing to my cultural and intellectual interests. If someone had tried to arouse my mathematical interest by presenting mathematics as a series of brilliant insights by a number of often fascinating people — unrolling it as a process of intellectual history, in other words — I should have responded as I did when I finally discovered the hidden charms of the history of mathematics. Watching glittering aspects of nature — in botany, zoology, geography, astronomy — being apprehended as an integral part of our historic growth became a different cry from those tightly locked subjects, complete with their own specialized terminologies, into which I had been so ineptly initiated at school.

The history of science also opened up a world of cultural flavor and color of which I had been unaware, adding a delightful dimension to my vision of the past and, in particular, of Medieval civilization. Two facts were at the root of this revealing insight — the actual vitality of science in past cultures and the fairly common antiscientific bias of the modern historian. The fascination with nature, the search for its theoretical and practical mastery, the intellectual or esthetic appeal of scientific inquiry, had been an integral part of virtually all earlier cultures, certainly of the Middle Ages, with each culture coloring science with its own characteristic hues. If historians were often unaware of this entire dimension, it was because the peculiar compartmentalizations of our modern two-culture civilization had produced this partial blindness.* Once I opened my mind to these realities, some exciting new colors, some earthy new flavors, and above all a new completeness had entered the picture.

I saw that the zest for technical invention and experimentation had caught on in the Medieval workshops. Speeding up a work process by installing a crank or a pulley presented an exciting challenge for Medieval craftsmen. Technical problems — how to hoist a huge marble slab and move it across town; how to make the sculptured folds in a mantle flow with a lifelike naturalness; how to place an enormous, conical cupola on top of an octagonal spire — preoccupied Renaissance artists just as the problems of statics inherent in a Gothic cathedral had been a matter of keen concern for Gothic builders.† What is more, scientific questions, instead of being shunned by writers or artists as belonging to an alien sphere, had moved to the foreground of a cultured person's

* There are a few eminent exceptions, the art historian Erwin Panofsky, the cultural historian Leonardo Oschki, and the historians of science George Sarton and Jacob Bronowski among them.
† There was, in fact, a committee of citizens from different walks of life deliberating for years on the problem of the cupola of the Florentine cathedral.

interests, so that Leonardo da Vinci's towering figure emerged as far less isolated (even though lonely), far more representative of the cultural currents of his time than I had suspected.

I had been used to thinking of technology as a purely utilitarian area, devoid of either intellectual or esthetic appeal. Its esthetic value was to me strictly negative — ruler-straight railroad tracks cutting a lovely landscape in two; rows of telegraph poles marring the innocent poetry of fields and forests. But Medieval technology proved to possess a surprising esthetic glow and to be an intimate part of the social landscape. A charming fresco showing the bustle of workmen raising a city wall in Siena with the help of a pulley offers as good a glimpse of daily life in the later Middle Ages as any peasant wedding or ice-skating scene by Pieter Breughel the Elder: pride in the newfangled gadgets was an authentic aspect of the Medieval mentality.

One gray summer morning in Paris I crossed the Seine into an amazingly drab part of the city. The Musée des Arts et Métiers was virtually deserted on this weekday morning. Walking around amongst the bizarre contraptions that make up its content — spiderlike monsters, medium-sized dragons devised for one or the other, often completely futile use, weird and clumsy forerunners of our modern industrial machinery predating the Industrial Revolution by hundreds of years — I realized that the growth of technology represented a significant chapter in the rise of the modern mind. Useful or not, "mainline" inventions or freakish deviants in the history of early engineering, these silent monsters were the tangible outgrowths of a kind of creative imagination that I happened to lack but that, nevertheless, embodied a high form of mental productivity.

It is from that experience, I think, that I find myself thrilled when I travel toward New York City through the final stretch of New Jersey, with its concentration of chemical industry — rows of giant chimneys, monstrous contrivances, barges under steam being loaded right on the factory grounds, huge billows of poisonous, multicolored vapors exhaled by the smokestacks, ramshackle industrial buildings still in active use. In a perfectly pleasurable way, Medieval culture has taught me to look at technological gadgets as visible expressions of the creative mind. Runaway from my own civilization that I basically am, I had to learn from the

Middle Ages that every technological object embodies a highly rational, admirably purpose-directed intelligence.

★ ★ ★

No doubt, the enrichment of my perceptions of Medieval culture was my most valuable gain from working in the history of science. Not only were Medieval people capable of amazingly trenchant scientific thought, like that brilliant fourteenth-century bishop of Lisieux, the cosmologer Nicole Oresme; what is more, nature played a far livelier role in Medieval literature and art than I had been aware of. Medieval poems and paintings were rich in natural references. Dante's *Divine Comedy* gave an authentic picture of the Medieval world-view by setting both worlds, this one and the one beyond, in an explicit cosmological context in keeping with the most advanced scientific thinking of his time, thereby sharply enhancing the poem's realistic effect. Gothic cathedrals were filled with reflections of the surrounding flora and fauna, the seasons, the farmer's annual cycle. The cathedral itself in all its mystic grandeur had to be understood as a reverent replica of the universe — the divine universe, to be sure, but one that was identical with, or rather manifested in, the natural cosmos.

Above all, I began to sense an integral unity between the Medieval and Renaissance and the modern experience of science: across the differences of the cultures and the boundaries of time, here was the same human mind struggling to grasp nature's laws and enjoying her dual challenge to the intellect and the senses. Leonardo had marveled *for us* at mountains and rock formations, trying to develop a concept of the geological structure of the earth. I cannot look at even an ordinary fountain without recalling Leonardo's drawings exploring the motion of water in its dizzying varieties. Leonardo was observing and exploring for us, epitomizing the questions — and tentative answers — of the preceding four centuries. The microscopic vision of modern science would have been inconceivable without the rigorous visual training of Renaissance art and the minute observation of natural detail that went with it.

One summer while I was working on this book, my wife and I were in the south of France and decided to take a side trip to Fontaine-de-Vaucluse, where Petrarch had his house and wrote his poems to the lovely, swift-flowing little Sorgue River. It turned out to be an enchanting afternoon — and an unexpected lesson in the meaning of Renaissance humanism. From Petrarch's fourteenth-century cottage, built of rough-hewn stones, a flight of steps led into an overgrown little garden filled with all sorts of herbs and medicinal plants, as well as purely decorative bushes and flowers, crowding in on the narrow footpaths. All of a sudden the formidable figure of the "first humanist," the great pioneer who had changed literature into a medium of personal expression about the joys of the ancient world, about scenic beauty and earthly love, appeared reduced to very human proportions indeed. This was the little garden he had cultivated, presumably every single day during the sixteen years he had lived here. The acceptance of nature, so vital at that historic point, had not been an issue in any big ideological program, nor simply an esoteric theme for his poetry. He had diligently worked his garden like any good Medieval monk and with his care had created this enchanting patch. His love of nature had merely been an extension of his loving care for this lay version of a cloister garden, redolent with its sweet and bitter scents.

As the historical landscape unfolded before my mind, I began to make out certain pervasive features. I could see the close link between the rise of modern science and the whole multilevel process of human liberation that we call the Renaissance. I thought I could even form a much clearer idea of the Renaissance itself than I had possessed before, or, it seemed to me, than we generally have, without the dimension of early science. From the respectable library of specialized books that has been compiled over the past fifty or sixty years, a picture of the evolution of Medieval and early modern science seemed to emerge that was no longer purely specialized but fitted into my general concept of history, complementing it in some telling respects. Science became an integral aspect of human activities during the rise of the modern world, a major liberating force, and, by becoming more human, a part of our own earlier development, our extended experience. We were that much richer and freer for having shared in the exploration of nature, all the way from Albertus Magnus and Roger Bacon to Leonardo da Vinci. Our common background had been that much more vibrant and colorful.

But not everything I learned was a source of vicarious pride or pleasure. Our age, the "age of science," when seen across a distance of eight hundred years, took on a startling, in some ways terrifying shape. Values we take for granted, on the order of absolute truths, turned out to be the result of historical accident, or at any rate of a chain of events having little enough to do with the true worth of our most cherished cultural axioms. We ourselves, part and parcel as well as products of modern Western culture, suddenly began to look considerably less sophisticated than I had assumed — no longer the highest personification of the human mind but objects of a complex historical process of which we did not even appear to be aware. Historical forces stemming from a distant past appeared to have generated the scientific and technological superiority of the modern West, not any act of matchless human wisdom. Were we any less the victims of inherited cultural prejudices than the people of the Middle Ages or Renaissance? Perhaps it was our *hubris* in confusing our intrinsic cultural biases with absolute truths that had stunted our critical detachment or muted our elementary moral revulsion in the face of the devastating impact of science on our lives — its share in the destructive potential of military technology, its increasing threat to our natural environment and cultural heritage; that is to say, to everything that makes life on this earth beautiful. Perhaps seeing science for the historical phenomenon it is, one of many types of specialized activity around which the human race has built a culture in the course of its history rather than the epitome of human wisdom, may help us to regain a sense of balance.

* * *

I have chosen to treat these aspects more explicitly in an epilogue, as afterthoughts to the discussion of the actual developments. In surveying these, my aim was to convey a lively, all-around picture of the major currents and their links with the surrounding cultural history, rather than to present any firsthand research of my own. Although my longstanding involvement with the subject has prompted some scholarly papers (and though the first and, to some extent, the final chapter are based on my personal studies), I decided in essence to write what is called a popular history, from a feeling that the subject is worth being opened up to the general reader on a generous scale, for its contribution to our general historical picture. By the same token, the book does not claim to be inclusive of all the important names or achievements of science in the Middle Ages and Renaissance. My emphasis has necessarily involved a selective process; I have chosen whatever detail seemed to be most characteristic or colorful, or provided the most telling illustration of a major scientific development.

* * *

In contemplating the expressions of gratitude customary in completing a book, I am reminded, first of all, of the actor Zero Mostel, who, on receiving the Tony Award for his performance in *Fiddler on the Roof*, got up before the solemn audience of theatrical celebrities and, to everyone's shock, in a mixture of Hebrew and Yiddish, thanked the Good Lord. Since "Fate" or "Fortune" plays a rather significant role in these pages — as, indeed, it did during the Renaissance — I should say that I do feel grateful to a fate that has permitted me to work on and complete this book on top of a fairly demanding academic career, as well as my scholarly work.

I am grateful to my friend Kenneth Heuer, a brilliant editor, lecturer, and writer on history of science themes, for spawning the idea of this book. My gratitude to him is even livelier for his reading the manuscript and encouraging and helping me practically toward getting it published when I showed it to him after all those many years. I am grateful to my editor at Houghton Mifflin, Ms. Anita McClellan, who contributed her final, decisive share in letting the manuscript mature into a printed book and did so with refreshing sympathetic tact and personal understanding. I am also indebted to my friends and fellow scholars John Parker, Norman Thrower, and Douglas Marshall, for reading the manuscript at an early stage and saying some encouraging things about it. And to a long line of charming young ladies who fought their way through the pathless jungles of my handwritten manuscript, and patiently and efficiently translated it into a presentable typescript. (Most consistent, most patient, and invariably most encouraging of them all, Ms. Constance Schwartz; and Mrs. Aime Andra, who, with her remarkable cooperativeness, persisted in seeing it through the final stages.) My thanks to Ms. Susan Carr, who helped me in assembling the illustrations, and did so with unfailing good taste, great resourcefulness, and infectious enthusiasm.

My wife, Helga, provided the ideal atmosphere in which such a book had to be written. With the right mixture of loving patience and ambitious impatience, never-flagging interest in listening to my many problems as I went on, and undiminished faith in its ultimate success, she makes me realize at the very end that, whether consciously or not, I have written this book for her.

Dawn of Modern Science

Brunelleschi's famous, widely visible cupola of the cathedral of Florence, topped by the lantern built by Michelozzo after Brunelleschi's design, which Toscanelli used for his observations of the path of the sun.

ONE

The Idea of the Earth in Renaissance Florence

O N J U N E 24, 1474, an old man — one of the most famous scientists of his time — was sitting in his study in Florence writing a letter to a friend. He was a somewhat tired old man, in his seventy-seventh or seventy-eighth year, and the things he was writing about frankly no longer interested him very much. Above the cathedral of Florence, in the lantern that Michelozzo the architect had placed atop Brunelleschi's bright orange cupola, he had set up an observatory a few years before. From up there, the panorama of orange rooftops and narrow streets and churches (and, farther out, the fields and villages and green hills of the Arno Valley) beneath his feet, he was making his observations of the path of the sun, using an elementary measuring instrument of his own devising.*

* Toscanelli's measuring instrument, called a gnomon, was famous in his time for its superior accuracy and praised by contemporary astronomers. Its purpose was to obtain exact data on the path of the sun, presumably as a frame of reference for a whole set of fresh astronomical observations, including the inclination of the ecliptic, as part of a critical re-examination of the known astronomical data, laid down in the so-called astronomical tables (see Biographical Note, p. 256). The gnomon consisted of two parts, one a heavy bronze plate that was embedded in the masonry beneath the south window of the lantern and contained a circular opening through which the sunrays could enter; and the other, a sundial that was located in the cathedral floor. Toscanelli must have conducted his observations essentially at the sundial (parts of which can still be seen in the center of the floor of the first chapel on the right of the left transept, which contains Michelangelo's Florentine *Pietà*). But the proper functioning of the gnomon probably required his frequent presence also inside the lantern, e.g., on cloudy days when the overcast sky called for closing off the lantern's windows,

Those were the things that intrigued him in his later years. They were advanced questions, questions into which only the vanguard of scientists in his time were probing: What was the exact relation of earth and sun? How did the movements of the sun appear when measured with mathematical accuracy? Precisely what were the workings of the solar system? About half a century later, these questions were to lead to a complete revision of the whole system of the universe, with Copernicus' theory of a sun-centered cosmos. In his lonely observations above the din of the streets of Florence, the aging scientist, Paolo Toscanelli, was in reality one of a handful of pioneers who were beginning to subject the traditional planetary system to a critical re-examination, men like the great cardinal Nicholas of Cusa (who had died ten years before), or his fellow German Georg Peurbach, and the latter's student who had brilliantly carried on Peurbach's work, Johannes Müller of Königsberg (who in the humanist fashion of the time had Latinized the name of his Franconian native city, calling himself Regiomontanus). This was the company in which Toscanelli was now moving in his intellectual pursuits, and these interests really marked the most advanced problems of science at that time.

A few days earlier he had received a letter from the distant city of Lisbon, from the hand of an old friend, Fernão Martins, now canon of the Lisbon cathedral. Martins had written in the name of no less exalted a personage than His Most Serene Highness Afonso V, king of Portugal. Unfortunately, the problem about which the canon and His Majesty were inquiring happened to concern a subject that, though it had interested Toscanelli well over a generation ago, right now was nothing short of a bothersome interruption of his astronomical observations.

What a capricious lady, fate! Here he was sitting before his desk, grudgingly writing out his explanations to satisfy the curiosity of

so as to obtain a more sharply focused beam permitting the exact measuring of its angle on the dial. Since Toscanelli's letter was written a few days after the summer solstice and he was especially eager to determine the location of the sun at that point, i.e., at its greatest distance (north or south) from the equator, it is safe to assume that he had to summarize his geographic theories at a time of particular absorption in his astronomical interests.

The construction of the lantern, which presented especially demanding static problems in which Toscanelli may have advised the architects, was begun in 1446 and directed by Michelozzo, following Brunelleschi's design.

Paolo del Pozzo Toscanelli (1397–1482), respected by his contemporaries as physician, astronomer, geographer, and creator of the modern concept of the globe.

the Most Serene King of Portugal; expounding them clearly so that, as he wrote, "the slightly learned" might understand, but with a certain tightness of phrasing that betrayed his fading interest in these ideas of long ago. Writing in his old man's humanist hand, mouth disdainfully curved downward below the enormous aquiline nose, a huge turban swathed around his head which he liked to affect in tribute to the Arab contribution to science; eager to get the letter, and the chart he would have to send along as an illustration, over and done with and get back to his little observatory high up on top of the cupola. (He made some excuses for not troubling to construct a globe, which he admitted would have made it all far more "evident to the eye.")

The letter has been lost, yet fate — or the tortuous ways of fame and accident that we call history — has been pleased to play her tricks on the old scientist and his rather casual letter. Some time later someone else, again writing from Lisbon, once more interrupted his studies with a request about Toscanelli's youthful geographic theories regarding the nature of the globe. This time, the old man merely took a copy of his letter to the canon and, under a few courteous lines, mailed it to his correspondent, a young man by the name of Christopher Columbus, whom he mistook for a Portuguese. He included a copy of the sailing chart he had sent to Martins.

In all likelihood the destructive forces to which documents and archives are exposed over long stretches of time would have devoured the second letter as they devoured the first, together with its chart (as well as the chart he sent to the young Columbus). But since Columbus finally emerged in a blaze of glory for having put Toscanelli's theories to the test and demonstrated their essential accuracy, that copy of the letter which the old man sent to him happens to have been preserved among Columbus' records. And that is why history remembers Paolo Toscanelli's name for the copy of a letter he had tossed off, in his old age, long after the subject ceased to hold a central place within his thinking. He is not remembered for the great cosmic questions to which he seems to have devoted most of his life's work and to which he may have made a major contribution.

The ravages of time have eaten away Toscanelli's astronomical writings, with the exception of a single manuscript that now rests in a Florentine archive, where it has survived the latest flood. Most of his other scientific studies have disappeared, as have his copious geographic notes, collected off and on over a long and extraordinarily active life. That brief, belated statement in his letter is all we know of Toscanelli's geographic concepts — and, for that matter, represents the only authoritative statement we possess of the ideas of the early Renaissance concerning the shape and nature of the earth.

Yet for all its casualness, for all those whims of fate, it is a highly important historical document: it is the first known statement of the modern concept of our earthly home.

* * *

Outside Toscanelli's study, the city was teeming with the rhythm of crafts and trades and the pulsebeat of its own animal vitality — as it had been teeming for the past two hundred years, and is today.

Florence in the 1470s was a city of vibrant energy and of colorful contrasts. Renaissance art was reaching a peak of perfection just then. The art of those years has left us a record of gratified pleasures, the leisurely contemplation of the moment. Little seems to have mattered to the artists and patrons of the 1470s other than what the senses perceived, what the eye to its delight could absorb.

Art mirrored the attitude of the city's leisure class, which had grown rich while Florence was Europe's leading commercial metropolis. Yet signs of decadence were already in the air — disturbing changes in the patterns of European trade, symptoms of business decline, rumblings of popular discontent. Money, which used to flow into business expansion, was now more and more channeled into art — a pleasant and rewarding enough diversion, but a symptom of economic stagnation nevertheless, which must have borne heavily on the working classes. The sophisticated elite around Lorenzo Medici, "Lorenzo the Magnificent" — the young man who five years before had assumed power over the city — developed a cult devoted to the sheer beauty of life, and ignored the warning signals. No doubt Lorenzo was magnificent — and so was the style of life he created around his person. Carefree, lavish to a fault, marvelously convivial, a serene little smile playing on his powerful features, he did his best to turn life in Florence into a great, high-spirited, never-ending festival. And for a brief instant — and for those privileged enough to enjoy it — he succeeded miraculously. In the memory of the West, the years under Lorenzo live on as one of the rare moments in the passage of time when humanity rose to the full enjoyment of living. Short-lived, confined to a small upper crust, not without occasional discords and

Lorenzo Medici, the "Magnificent" (1449–92; from a portrait by Agnolo Bron-
zino), who personified the peak of the Florentine Renaissance.

signs of self-conscious guilt feelings though it was, Lorenzo's Renaissance stands as a model of a culture inspired by the enjoyment of the earth and its beauty.

In a few ringing Italian lines, Lorenzo had flung his credo into the enamel brilliance of the Tuscan sky:

> Quant'è bella giovinezza
> Che si fugge tuttavia!
> Chi vuol esser lieto, sia:
> Di doman non c'è certezza!

Youth is beautiful yet slips away. Come on, let us be happy. Who knows what tomorrow will bring?

Here was a credo based on the abiding beauty of the moment, tinged — and perhaps deepened — by a poetic awareness of its transience.

Lorenzo's motto seemed to echo through the art of those years. Earlier generations had struggled with the laws of perspective, anatomy, and movement. As a younger man, Toscanelli had advised the painters in the mathematical principles governing visual perception. But now a new crop of artists was holding sway over the city. They had easily mastered the new techniques during their apprenticeship and were now applying them in their studios to leisurely views of the life of the upper classes, the common people, or the poetry of the countryside. They might use a conventional subject from the religious stories of the past — the Adoration of the Magi, the Birth of the Virgin, the Birth of Christ — and would then people the painting with well-known members of the Medici family and their friends, with their own portraits, or with the faces of ordinary men and women from the Florentine streets. Or they would offer a loving view of the maternity room in one of the private houses of the Florentine upper class (all drawn in perfect perspective), with elegant ladies in their gorgeous dresses trying to catch a glimpse of the new-born, future Mother of Christ. Wherever they looked, whatever they saw, these artists of the new generation were feasting their eyes and inviting the public to share their pleasure. The world was lovely, life in Florence a never-ceasing wellspring of poetry — if one only knew how to see it.

Andrea del Verrocchio's LITTLE BOY WITH DOLPHIN. Florence, courtyard of the Palazzo Vecchio (circa 1470).

Opposite: Elegant Florentine ladies crowding into a maternity room. Detail from Domenico del Ghirlandaio's fresco, BIRTH OF THE VIRGIN, Florence, Santa Maria Novella, Cappella Maggiore ("Tornabuoni Chapel"), 1485–90.

The sculptor Verrocchio completed such works as the *Little Boy with the Dolphin,* an enchanting symbol of the quiet charm of those years. Standing in the center of the cool Renaissance courtyard of the Palazzo Vecchio, Florence's Medieval City Hall, a steady jet stream of water spouting from the dolphin's mouth, the sculpture seems like a radiant little smile greeting us out of the 1470s. The young Michelangelo began to study sculpture in Lorenzo Medici's garden.* It was still under Lorenzo's rule that

* Vasari's appealing story — all the more appealing because the garden may be easily inspected through a wrought-iron fence from Via San Gallo — has been questioned by E. H. Gombrich ("The Early Medici as Patrons of Art," *Norm and Form,* London and New York, second edition, 1971, page 56). But even Gombrich's cautious probing admits to some connection between Michelangelo's studies and the Medici gardens.

Botticelli painted his love goddess, Venus, her naked limbs rising from the sea. In the luminous quiet of the Tuscan night, philosophers and dilettantes would gather to discuss Plato at one of Lorenzo's country estates.

Still, there were discords amidst this symphony of rarefied pleasures. While Lorenzo lavishly supported artists and humanists, the sprawling international banking house his grandfather Cosimo had built declined, and two important branches, at Bruges and London, folded in bankruptcy. What was more, the people of Florence began to grumble against Lorenzo's rule. Grandfather Cosimo had set up the Medici regime over the city with infinite caution, governing by a subtle system of indirect means, anxiously avoiding any ostentation that might anger the people. Lorenzo did not seem to care what opposition he might provoke as he displayed his power more and more frivolously. Feeling far too secure in the power structure he had not built himself, he and his band of reveling cronies would noisily disturb the sleep of the good Florentines with their singing late in the night, heedless of the mounting anger behind shuttered windows and tightly barred doors.

There were deeper undercurrents of popular discontent besides the growing outrage over the frivolity of Lorenzo's clique — resentment of his financial extravagance, of his callousness toward the elementary needs of the people, his more and more overtly dictatorial rule — all creating a sullen atmosphere of indignation over the whole hedonistic philosophy of Renaissance culture sponsored by the Medici family. As later events were to show, the common people in their workshops must for a long time have felt bitterly critical of all that fancy paganism that went under Lorenzo's name. In 1478, a violent outbreak killed Lorenzo's younger, dearly loved brother Giuliano and had to be bloodily suppressed. That was just a mild forewarning of what was to come. A few years later a somber and fanatic Dominican friar, Girolamo Savonarola, began to thunder Sunday after Sunday from the pulpit of the cathedral against the sinful life of the upper classes, and the people crowded into the *duomo* to hear him, week after week. At

Opposite: Girolamo Savonarola, from Raphael's fresco DISPUTA. Rome, Vatican, Stanza della Segnatura (1509).

last, two years after Lorenzo's premature death, a popular uprising, inspired by Savonarola, swept away the Medici rule. Bonfires were lit and consumed many a Renaissance treasure that remorseful citizens and their ladies, in a great wave of public penitence, threw into the flames. We do not know how much great art may have been destroyed by this holocaust.

Yet in 1474 the whole, marvelous celebration was still in full swing. Discontent had not yet broken the surface. The great economic changes had not yet transformed Western European society, leaving Italy behind, as they would during the following century. Florence was still the leading emporium of trade, the center of early industry. Despite the political resentment, the craftsmen went calmly about their daily work in their workshops, perfecting tools and elementary mechanical equipment to mark important technological strides. The irrepressible vitality of the city provided its tonic day after day. In fact, some of the unique dynamism behind the evolution of fifteenth-century art, its intrinsic intellectual energies that were frequently bordering on scientific problems and at last found their fullest expression in the genius of Leonardo da Vinci, may well have been nourished by those undercurrents of tension and discontent. Significant contributions to Renaissance art and science came from sectors outside the Medici elite.

Still, for most of the artists, and the patrons for whom they worked, those were years devoted to the joys of pure vision. The laws had been formulated by which the eye assimilates the rhythmic patterns of a street, the harmonious order of a room, the movements of a crowd, the beauty of the human body. The painters of the new generation — Ghirlandaio, Pollaiuolo, Filippo Lippi, the young Botticelli — were concentrating their technical skill to bring out the inner beauty of it all, our earthly environment, their enthusiasm entering into every nook and cranny, like the good sunlight itself.

Quiet fountains were splashing in the patios of the *palazzi*. The streets were hushed in the noonday heat, ebullient in the morning, noisy and crowded in the afternoon and evening, till the night brought its serene silence. The Renaissance had added some lovely houses and churches to the town, but otherwise it had all been there for hundreds of years, the life of the city with its unending subject matter for the artist. Why had art only now turned its full attention to the fascinations of the environment?

*　　*　　*

True, it had all been there for centuries. Yet if art is an expression of our visual awareness, it seems equally true that people had not really taken a good look at their daily surroundings for a very long time. They had not really "seen" that perennial show of crowds milling around among narrow streets, swapping jokes, gesticulating, flirting, driving bargains. They had not seen the beauty of shade and sunlight, of brook and foliage, of sky and cloud. To have an object in front of one's eyes does not necessarily mean that one perceives it, let alone experiences its emotional impact. Our experience is conditioned by a multitude of subtle elements, and many of these stem from our cultural situation — that is to say, they are caused by historical factors.

In the consciousness of the West something like a gradual "discovery of the earth" had, in fact, taken place during the three-hundred-year span between the twelfth and the fifteenth centuries. Although art and literature had supplied the obvious media for expressing the perception as it gradually focused on the earth, science had registered the intellectual steps in this complex cultural and psychological process.

For close to a thousand years the human imagination had been steeped in the contemplation of otherworldly orbits. By the twelfth century the change had set in with refreshing suddenness — and yet from then on the adjustment had, on the whole, been hesitant and slow. The façades of Medieval cathedrals began to blossom forth with sculpture showing trees and human faces and animals. Frescoes depicting pious subjects along the walls of a somber church began to mirror real streets and houses, actual people. The covers of handwritten manuscript volumes and a very few paintings began to display glimpses of natural scenery, some childlike rendering of a garden, the clumsy outlines of a town situated on the shores of the sea or a lake. Dante had interrupted his voyage through a world after death with a sudden, radiant glimpse of the Mediterranean at daybreak. Giovanni Boccaccio spun his tales about people in all their native humanity, amidst the hustle and

The transition from traditional Medieval to early Renaissance outlook, as expressed in art, tended to occur rather abruptly. The CRUCIFIXION by Giotto (*above*) in the church of Santa Maria Novella, Florence (circa 1300), is still a monument to traditional other-worldly spirituality. By contrast, his DEATH OF ST. FRANCIS (*below*), in the Bardi Chapel of Santa Croce, Florence (circa 1320), portrays intense religious feeling in the human terms of St. Francis' brethren grieving over his sudden death.

bustle of a real Italian city. Scientists became accustomed to look closely at details of nature, such as a cloud formation or a leaf of grass.

An immensely subtle process was taking place in the vision and consciousness of the Western world, but for all that its basis was perfectly real. Its roots were in the reality of a changing social life, produced by the first stirrings of the early capitalistic system. Yet the changing vision occurred within the subtle recesses of the mind. It involved the responses of the Medieval mind to the changes that were taking place on the economic and social scene and, still more subtly, the way these responses were determined by the prevailing cultural tradition, which was in essence the Medieval tradition of spiritualized otherworldliness. It was within this same area of delicate interactions between fresh responses to a changing social reality and traditional otherworldly, eminently abstract attitudes that the birth of Western science occurred.

The fifteenth century saw the triumphant consummation of this process. If until then glimpses of the human and natural environment had been no more than intermittent, if the mental adjustment had been nothing more than gradual and the mind's "return to earth" on the whole reluctant and slow, the Renaissance received its full significance from the joyous completion of the discovery of the earth. For all its naïve exuberance, that leisurely exploration of the daily environment that inspired the painters of the 1470s was really the fruit of a long and complex adjustment of the Western mind. Earth had at last been accepted, those deep-rooted otherworldly traditions effectively assimilated into the new positive view of life in this world.

The earth had dawned on Renaissance Florence in the minute details of artistic observation that the laws of perspective and anatomy had opened up. And the earth had emerged in scientific detail as a geographic vision to Paolo Toscanelli and his friends, very much as part of the process of discovery.

* * *

We may think that the earth must always have looked just about the same to its human inhabitants. But it did not. Earlier misconceptions were not so much that the earth is a disk or a rectangle, rather than a sphere. That was the way uneducated people may have seen the shape of the earth, and at certain times — as in the early Middle Ages, when scientific thought was weak — such primitive ideas were even generally accepted. Serious scientists, however, had thought of the earth as a globe at least since the days of ancient Greece.

The major misconception ruling even serious scientific thought for some three or four thousand years concerned the extent of the human habitat: the "habitable earth" was supposed to cover only a fraction of the globe. The rest was viewed as something like an inaccessible outer sphere. In the thinking of the West, only the European continent and its surrounding areas formed the *orbis terrarum*, the "sphere of land," or habitable sphere. Its scope extended as far as the geographic knowledge of the people inhabiting the Western world. On the old maps we can see how that knowledge got hazy when it came to the Scandinavian North. India up to the Ganges was known to the West mostly from Alexander's expedition, and that was as far as Asia appeared on the maps for something like the following eighteen centuries. Beyond the Ganges, and toward the North Asiatic steppes, the habitable earth faded once more into the shadows. Of the African continent only a sliver was known to the ancients, where it borders on the Mediterranean. Farther to the south, Africa lost itself in the desert sands and was at last cut off by the Ocean Sea. To the east, Africa ("Lybia," the Romans liked to call it) was linked by a massive land bridge to Asia, locking the Indian Ocean and its spice-growing island world in its grip.

This, then, was the world the European mind lived in from ancient times through the Middle Ages into the days of the Renaissance: a solid complex of land consisting of the three known continents (or really only portions of them), a world detailed and fairly accurate in its features around the familiar locale but, farther away, fading into the shadows — the *orbis terrarum*, the *oikoumene* of the Greeks, on all sides enclosed by the great dark and unknown Ocean Sea. Times of little scientific culture were to play their games with this picture (even after the spherical shape of the earth had been established by Pythagoras toward the end of the sixth century before Christ), flattening the globe into a disk or a rectangle, dividing the earth into four equal segments, setting

The EXPULSION FROM PARADISE by Giovanni di Paolo (circa 1445), originally painted for a church in Siena, now in the Herbert Lehman Collection of the Metropolitan Museum in New York, incorporates a contemporary map surrounded by the spheres of the Aristotelian universe within the representation of a dramatic religious theme. (The map, a fairly crude example of Renaissance cartography, shows east — rather than north — at the top in the traditional Medieval manner, with Paradise, including the four rivers that according to Genesis originate in Eden, located in the Far East.)

the Christian hell in one corner and earthly paradise, the Garden of Eden, in another. Yet even in ages of intense intellectual vitality, people like Plato or Dante, who were in full command of the scientific culture of their times, believed to be forever confined to a relatively small portion of the earth behind that huge watery prison wall formed by the Ocean Sea that enclosed the habitable sphere of land on all sides.

Water, then, represented the unknown — and impenetrable — element, the inaccessible "outer space" confining the human habitat and movement. When the age of discoveries succeeded at last in surmounting this wall, "space" loomed as the next enclosure — one that the twentieth century is now beginning to pierce.

* * *

It seems strange to realize that through thousands of years people should have thought of their earth in such severely limited terms.* Such concepts reflect, of course, a limitation of the actual geographic knowledge, but the fears of an unknown element occupying most of the globe suggest other implications, psychological and even philosophical. They suggest something of the degree to which the human race still felt a stranger on the face of the earth, over surprisingly long stretches of history, and how certain cultural contexts could renew that sense of alienation.

At the dawn of civilization, geographic concepts had clearly been a part of the ruling "animistic" view of the world, which suggests how early civilizations in their native habitat felt surrounded by a world of spirits embodying the forces of nature. Greek culture boldly pioneered the exploration of the earthly environment beyond the familiar bounds (after the groping beginnings of the Egyptians, Phoenicians, and the seagoing people of Crete). The Greeks also initiated a vigorous trend of descriptive — as well as speculative — geography, which at last, in the Hellenistic age, led to a vogue of authentic geographic science.

Yet as the otherworldly philosophy of the emerging Middle Ages, after the decline of the classical world, was lifting the mind to its esoteric heights, the shadows of ancient fears once more fell across the globe. To the Middle Ages the earth seemed host to terrors so vastly beyond the familiar experience that they staggered the imagination. The fears about that watery sphere assumed nightmarish shapes from Frankish times to the age of the great explorations, harassing peaceful citizens in their sleep from the seaports to the cozy inland towns: the slimy sea serpent, sea monsters of every description, fast-ducking islands that were really the backs of whales, dead men condemned to hell whom the Ocean was believed to harbor at its bottom and whom fishermen would find sprawled on the beach in the morning after a storm.

Yet nightmares were not all that the Ocean inspired. The Medieval mind also projected its hopes of a better life, its social and

* The first known maps reflecting these concepts were in the form of clay tablets, now in the British Museum, made in Akkad or Sumer between 2500 and 2100 B.C., showing lower Mesopotamia — the known world of those civilizations — surrounded by the "Ocean River."

A map from the early Middle Ages (circa 900 A.D.): east is again at the top in this schematic presentation, with Adam and Eve (and the snake) indicating the site of Paradise; Europe is at the lower left, separated from Africa by the Mediterranean. The Ocean with its islands (some legendary, some nameless) appears as a circular river, representing the "sphere" of water surrounding the "sphere" of the earth.

personal utopias against the horizon of the element. Legends peopled the unknown sphere with land formations inhabited by beings resembling the human species yet living a better, freer, more wisely ordered life: the "Fortunate Isles," the "Earthly Paradise," the "Island of the Seven Cities," the "Isle of Men," the "Isle of Women" (or of the "Amazons") — names that were later perpetuated on the maps of the New World, like that wondrous isle of "Brazil," or the Amazonas River.

Beyond the habitable sphere of land, then, even beyond the vast expanse of the Ocean, were races of "antipodes," wise and kind beings living in the perennial contemplation of a blissful scenery;* or races of women indulging in sexual practices that clearly were but a wish fulfillment of a more inhibited everyday world. The geographic imagination of the Middle Ages and Renaissance did not distinguish between fantasy and empirical fact. The legendary islands duly appeared on the maps — and were only gradually replaced by the actual findings during the age of discoveries.

At bottom, what really troubled the mind for thousands of years was the question the twentieth century now asks about space: may not the unknown sphere nourish familiar forms of life? Not unlike our space speculations (in which modern science fiction takes on the role of the Medieval legend in giving vent to rank imagination), a trend of positive scientific thought had long answered that question in the affirmative, simply on grounds of likelihood. Already, during the time of Christ, the geographer Strabo had written: "It is conceivable that in the same temperate zone — i.e., which we inhabit — there are actually two inhabited worlds, perhaps even more, and particularly in the proximity of the parallel through Athens in the region of the Atlantic Ocean." Before him, Plato had already speculated about a lost transoceanic continent, the famous Atlantis. From then on, through Roman times, through Medieval fantasies about Oceanic islands, imagination and rational thought had conjured up all kinds of archetypic forms of the undiscovered New World. Toward the end of the fifteenth century, around the time Toscanelli was writing his letter, the belief in the existence of another "world," a "fourth continent" somewhere out in the Ocean sphere, had assumed the dimensions of certainty.

Strangely, for something like two thousand years the undiscovered American continent had reflected its presence in the speculative thought of the Old World. The mind had created the New World, by bits and pieces or as a coherent entity, long before explorers located its actual shape by their struggles with the elements.

* * *

* This age-old Medieval myth became a favorite dream of the Renaissance, finding its classical expression in Thomas More's *Utopia*. It lived on in European literature till the eighteenth and nineteenth centuries, in the Romantic idea of the "noble savage."

Paolo Toscanelli's contribution to modern geography consisted of two equally crucial feats: he proclaimed the navigability of the Ocean, thereby establishing the whole earth as the human domain; and, through this theoretical achievement, he unlocked the Ocean for the discovery of the New World. There had been poets within that long speculative tradition who had anticipated the modern globe. The Roman poet Seneca had predicted an age when the chains of the Ocean would be broken and the earth at last lie wide open for human exploration.*

Now that age was here. If the practical breakthrough was achieved by the great discoveries from the early Portuguese to Columbus, Vespucci, and Magellan, the credit for the intellectual feat that made that liberation possible goes largely to Toscanelli. By the force of careful, vigorous thought, he worked out a cogent concept of the earth in which the whole Ocean was accessible to people and their ships, and the solid land portions of the whole earth inhabitable, East as well as West, the Northern and the Southern Hemispheres.

How did Toscanelli arrive at this revolutionary vision? He was actually not the sole author of the modern concept of the earth. For almost a generation during the earlier part of his life, somewhere between 1410 or '15 and 1440, he had been engaged in intense discussions about geographic problems with a group of Florentine humanists, among whom he was clearly the leading scientific mind. Not that this sort of informal teamwork was unusual for the advances of science during the Renaissance. The laws of perspective and anatomy were largely worked out through discussions among groups of artists. The invention of movable print developed out of the experiments of a number of craftsmen exchanging experiences between their workshops. Each one of these discoveries took several decades to mature, and, in a final spurt of energy, was completed during the 1430s — the time when the new mental map of the earth was taking its shape.

* The passage is amazingly prophetic, suggesting the impressive scope of ancient geographic thought: *". . . venient annis / Saecula seris quibus oceanus / Vincula rerum laxet: et ingens / Pateat tellus: Typhis novos / Detegat orbes: nec sit terris / Ultima Thule."* (There will come an age in some distant future when the Ocean shall loosen its shackles and the earth shall lie wide open; and Typhis [Jason's guide, hence an explorer-sailor] shall discover a new world. And no longer shall there be an end [Ultima Thule] to the earth.)
Seneca, *Medea,* Act II, Scene 2, line 371

The year 1410 had seen the publication of a book that in one big leap updated the geographic thinking of the Western world, raising it to the most advanced level attained by antiquity. Almost overnight, with the appearance of the Latin translation of Ptolemy's *Geography*, a gap of some twelve hundred years was bridged, and geographic science could resume at the point where the ancient world had left off. A young Florentine, Jacopo Angelo de' Scarperia, had worked on the translation from the original Greek for close to five years, serving his age's dual curiosity about the classical past and the detailed features of the earthly environment. The publication caused an immediate and enormous stir. Informal discussion groups like that of Toscanelli and his humanist friends seem to have sprung up almost at once.

Ptolemy had several important things to teach his fifteenth-century readers. For one, he taught a lesson in the straight empirical approach to geographic information. The fog of legends lifted before his cool Hellenistic mind. He also taught how to "see" the earth, in every one of its segments, in correct mathematical terms. As visually oriented an age as the early Renaissance must have been fascinated by Ptolemy's chapters on the methods of spherical projection. (In fact, modern cartography — and the modern atlas — received its start from the discussions of the *Geography*.)

The critical problem of the Ocean sphere, as pictured by Claudius Ptolemaeus, did not involve any explicitly radical new concept. Ptolemy was too soberly empirical for that, too much averse to large and daring generalizations. But it was exciting to the fifteenth century precisely because he was so doggedly empirical and sober. Just about the same time the translation of the *Geography* appeared, the vivid geographic interest produced an extremely popular summary of the Medieval picture of the earth from the hand of a French cardinal, Pierre d'Ailly, the famous *Ymago Mundi*. Amidst his hodgepodge of Medieval lore the good cardinal staunchly perpetuated the idea of the surrounding Ocean River — in passing permitting himself a haughty little sneer at "certain modern philosophers" who were beginning to question that time-honored idea.

Ptolemy had it otherwise. His earth was free of nameless terrors or twilight zones, everywhere as open and inviting as one's neighbor's backyard — almost the wide-open earth that the poet Seneca envisioned. His habitable land mass was, in fact, remarkably spacious and detailed (the African continent extended as far as 15 degrees south). And even though Ptolemy did not explicitly say so, he clearly seemed to assume that the Ocean was simply another navigable waterway. On the face of his lucidly profiled earth there certainly was no room for such specters as the "torrid zone" — that molten mass which supposedly enveloped the entire Ocean below the equator like a giant furnace in which ships would catch fire and crews have their skins scorched black from the heat.

How much more than d'Ailly's musty old jumble shop Ptolemy's earth suited the fifteenth-century experience! A few years after the publication of the *Geography*, news began to reach Florence that the distant young nation of Portugal, located right at the rim of the Ocean Sea, had begun to send ships into the unexplored outer orbit. A young member of the Portuguese royal house, Prince Henry the Navigator, was masterminding these expeditions. In his portraits he looks out of a somberly carved Medieval face through remarkably bright modern eyes that are set on some distant future. Perhaps his origins — part English, part Portuguese — had something to do with his curiously blended looks and mentality.

News of the discoveries under "O Navegador" came thick and fast, giving Toscanelli and his friends some exciting new geographic insights, over and beyond the study of Ptolemy's text: a scientific expedition to the Canaries in 1416; colonization of Madeira in 1420; discovery of the Azores between 1427 and 1432; regular voyages down the west coast of Africa, culminating in the rounding of the dreaded Cape Bojador in 1434 and, consequently, entrance into what even the more factual-minded Arab geographers were still calling the "Sea of Darkness."

The eastern Atlantic was opening up. The Ocean proved to be navigable for carefully planned expeditions, even though these limited inroads meant as yet little more than dipping one's toe into the water. And Ptolemy was now proven right on at least one demonstrable fact: Africa evidently extended much farther south than the Medieval maps had suggested.

We are not exactly certain how such momentous new data were received and evaluated by Toscanelli's little group. Did the participants greet each other in the streets of Florence with the latest bit of information from Portugal? Did they call a special session in the Tuscan hills, at one of the abbeys they used for their meetings, to fit the latest discovery into their slowly growing picture puzzle of the world? Our fragmentary evidence permits us to catch some occasional snatches of the conversation, not the whole dialogue. But we do know that the Portuguese, who were also studying Ptolemy's book, were in frequent touch with the Florentines and, on occasion, would pump them for some significant bit of factual detail or geographic theory.

Portuguese caravels were beginning to plow Atlantic lanes and sail back into Lisbon's harbor, on the Tagus, with cargoes of strange fruit and, in time, ivory, gold, and African slaves. Gathered under the cool library vaults of an abbey, looking out into the Tuscan night, Paolo Toscanelli and his friends must have felt that the earth was expanding. Soon their horizon was to widen even more spectacularly.

* * *

Behind every creative feat lies a whole complex of motivations. The reasons that brought together a number of highly educated and alert men — some of them famous scholars of classical antiquity — to work out a new picture of the earth were in part extremely subtle, in part downright practical, even commercial.

The more subtle ones had to do with the climate of curiosity about the earthly environment that pervaded the Renaissance. During the early fifteenth century, even while these geographic discussions were going on, the avant-garde of Renaissance artists was concerned with the problem of people moving about on their earth. Sculptors like Ghiberti, Donatello, and Verrocchio created figures boldly standing free with no support other than their feet — no longer the anxious, wall-hugging sculptures of the Medieval tradition. Painters like Masaccio stunned their public with daring experiments in perspective, creating the illusion of having torn open the wall in back of their figures to provide a foreshortened glimpse into space. The flat, opaque background — as a rule, golden or black — of the traditional Medieval paintings had been torn down, as it were; a new dimension to move in had been opened.

For the frame of mind in which these experiments were under-taken (or received by the public), the geographic studies provided yet another extended dimension. Where the artists explored the scope of the immediate visual perception — and the problem of motion in art, of the human figure moving around in its imme-diate setting — the geographers were exploring the earth as a whole. For all the evident differences in their media and their sheer scope, both geographers and artists were giving their public detailed and, to the best of their knowledge, accurate observations about the nature and shape of the earth, at the same time convey-ing a feeling of heightened mobility on our native planet (which at least in the geographic studies led to spectacular practical re-sults). Artist and geographer were both engaged in a "conquest of the earth" — the one within the scope of the eye, the other as far as the human mind could range.

Nor should one overrate the barriers separating art from science during the Renaissance. Renaissance artists — not only Leonardo da Vinci — were excellent observers of nature (besides being well versed in the technological implications of their craft) and some-times excellent mathematicians. Their ultimate aim, the explora-tion of the earthly environment, seems often more intriguing to the creative Renaissance mind than the type of specialized activity by which it was achieved. The personal ideal of the age was, after all, *l'uomo universale* — universal, all-around man. The detailed geographic descriptions of familiar parts of Europe (especially of sections of Italy) that were a vogue during the Renaissance were really scientific companion pieces to the paintings and etchings of cities or of vistas of the country scene. Maps of cities like Florence, Venice, Genoa, Rome, carefully drawn in realistic relief, were usu-ally executed with remarkable artistry; they can hardly be dis-tinguished from the numerous Renaissance paintings featuring the detailed panorama of a town.

As the progress of the age of discoveries expanded the knowl-edge of the world, the same type of geographic literature and of artistic mapping came to extend more and more to the distant parts of the world. Geographic studies, then, were not only what we would take them to be, a specialized (and perhaps somewhat limited) branch of empirical science; they were an important facet of the major intellectual adventure of the time, with definite esthetic and even artistic implications.*

If geographic exploration was in this sense akin to the art of the Renaissance, practical motivations were also very much present and clearly had their share in the evolution of geographic thought. At least some of the geographers were involved in the spice-importing business. Toscanelli's family owned a respectable firm in the trade, and he was himself actively engaged in the business for years. Columbus, as a young man, was associated with the importing of spices (which was one of the reasons why he had written to Toscanelli, inquiring about his geographic theories). Portugal launched her expeditions very largely in the hope of finding a new route to the spice-producing areas of the East.

There had been some distressing upheavals in Asia that had caused serious dislocations along the traditional trade routes, over which the spices were carried to the West. Spices were not just a luxury item for the Medieval or Renaissance economy. In food they provided, besides seasoning, the main means of preservation over any period of time. In pharmaceutical medicine, spices simply took the place of the modern chemicals. Spices, in short, were indispensable for health and daily living. As a physician, Toscanelli was, of course, familiar with the medicinal uses of spices. As a member of an importing firm, he knew about the market problems growing out of the recent shipping difficulties.

Undoubtedly, interest in the import of spices added an element of practical urgency to what otherwise might have remained little more than an intellectual pastime — piecing together a more authentic picture of the earth. It also expanded the scope of the geographic studies. It meant, for one thing, that our Florentine geographers were eager to gain a clearer, more detailed picture of the Indonesian island world where most of the spices were grown — and therewith of the general geography of the Far East. It meant that they began to wonder about possible new routes for reaching

* Among the subtler motivations (to be discussed more in detail in Chapter Six) was the urge to assert "worldly" surroundings against the world-denying legacy of traditional Medieval culture, a basic Renaissance theme. Art often drew on the geographic imagination in this pursuit, by presenting the world of sense perception in the alluring colors and with the typical features of exotic countries.

that part of the world, which involved nothing less than a complete revision of the traditional concept of the globe.

Oddly, then, the final spur to the geographic thinking of the Toscanelli circle came in all likelihood from the import business. The geographers' minds presented a curious mixture indeed: the humanist scholar's thirst for the ancient knowledge; the trained scientist's systematic approach to a given problem, which had been developed over the past three hundred years in several areas, even though not markedly with regard to geographic studies (a training Toscanelli had acquired as a student at the famous School of Padua); the salty experiences of Portuguese sea captains and sailors, filteied through O Navegador's expedition headquarters at Sagres in the Portuguese South; and, as a last, exotic-flavored ingredient, the worldwide business thinking of the importer of spices from the Far East. This whole astounding mélange may not be considered the proper intellectual climate by a modern scientist, but it was germane to the far less specialized, buoyantly creative atmosphere of the Renaissance. What matters most, it produced results.

The fact is that Italian businessmen had been used to far-flung international transactions — and to the sort of global thinking that goes with worldwide trade — ever since the thirteenth century. Aside from Far Eastern imports, the big Florentine merchants had been trading in woolens and silks over an international market extending from England and Flanders all the way to North Africa and the Near East. Reports from their business agents in London, Bruges, and Damascus, Aleppo and Tunis, would reach the Florentine counting houses in a ceaseless flow, enabling the great merchants to keep abreast of the changing international market scene, to adjust their prices and sales, to mobilize funds or credits. Especially during the big boom of the early 1300s, Florentine businessmen had developed the managerial and banking techniques needed to reap the highest profits from this growing world market. In fact, it was at that time in Florence that the institutional basis for modern capitalism was built. It was at the beginning of this early capitalistic boom that the famous adventurer from Venice, Marco Polo — prototype of the wheeling-and-dealing big American businessman, as Eugene O'Neill has seen him —

came home from his years in the glittering East and, with the publication of his memoirs, exploded the geographic horizons of the Italian middle class.

Economic expansion had become somewhat slower, trading a bit less ebullient, when that first boom collapsed in a crash. True, Italy was still the great trading center (even though not for much longer), the mart of Europe from which the merchandise of the Orient — silks, tapestries, perfumes, refined steel products, and always, above all, spices — were shipped to other European countries. The major Florentine merchants (the Toscanelli family firm with its branch office in Pisa has to be counted among them) were still thinking in world-spanning terms, within the limits of the available geographic knowledge. But those upheavals in Asia, threatening enough to make European statesmen try to whip up a belated crusade against the advancing Turks, had upset the picture of a smoothly flowing Oriental commerce. And the first rumblings of profound changes in the European economy, shifting the center away from the Mediterranean toward the European North and Northwest, were already discernible to a sharp businessman's ear. A merchant's best thinking, the full weight of his knowledge, had to be brought to bear on the problem of finding a different route to "the places where the spices grow" — including the use of ancient texts, the most advanced scientific methods, and the most detailed data about faraway regions on which one could possibly put one's hands.

From a strict, "pure," scientific viewpoint, the commercial motive was certainly an ulterior concern. But science does not always advance for purely scientific reasons, nor in an atmosphere of abstract scientific thought. The new mentality that sprang up with the early capitalistic type of business operations had a great deal to do with the rise of Renaissance culture as a whole. It was particularly developed in Florence, with its leading position in international banking and trade. From the late thirteenth century, when Florence had emerged as a highly advanced, typical early capitalistic community, complete with the social tensions between management and labor and their conflicts over political power in the government of the town, social and economic factors had strongly influenced the cultural outlook. They were the spur of reality behind those subtle changes in the visual perception. The awareness of living in a dynamic new environment that received its lifeblood from the worldwide exchange of the treasures of the earth had turned the citizens of Florence against the world-denying Medieval tradition more fiercely than people anywhere else. By asserting the glory of life on earth in painting and sculpture, Florence was giving expression to a new style of life, the outlook and life style of the modern age.

Paolo Toscanelli and his friends were merely projecting the wide-open spirit of their city into their geographic studies. They were adding another brilliant facet to the earth-oriented culture of the Renaissance.

* * *

Although Toscanelli's geographic notes have not been preserved through all the vicissitudes of private repositories and public archives, we know at least that he collected them over a number of years — and may surmise that they included every scrap of interesting information he could possibly get hold of. One such fact-gathering episode he describes in his letter, a long interview he had with an emissary of the Great Khan about conditions in the man's country, the Mongol Empire of the Far East.

Around the same time, an opportunity for very extensive additions to that mental map he and his friends were piecing together presented itself in the person of Niccolò de'Conti, another Venetian who returned from the Far East and came to Florence in 1442. It seems that Conti was invited to attend several sessions of the Toscanelli circle, where he gave a remarkable report, an up-to-date, more factual version of Marco Polo's observations of almost a hundred and fifty years before. Later, a famous humanist, Poggio Bracciolini, published Conti's report among his own writings, probably from notes he had taken while Conti talked to the Toscanelli group.

The features of the Far East were moving into increasingly sharp focus: world maps produced in Italy about this time reveal the mapmakers' as yet fairly clumsy efforts to be as specific about these regions as they were about European geography (which they updated by adding the recent Portuguese discoveries in the eastern Atlantic). All this was fine for the progress of science, but it still left unsolved the question of how one could get there without using the old routes through the Near East and paying steep tolls to the sultan or else running into all kinds of Turkish chicanery. At last, the solution arrived in the guise of a highly remarkable man.

Georgios Gemistos Plethon's dramatic shadow has flickered across many an aspect of the intellectual life of the Renaissance.

Andrea Bianco's "world map" of 1436 is a great deal more detailed and sophisticated than the one in Giovanni di Paolo's roughly contemporary EXPULSION (p. 17). In the Medieval tradition, Bianco's map, too, has east at the top. However, instead of schematic subdivisions and Biblical tales it contains a very explicit (if not exactly accurate) idea of the Far Eastern island world and a fairly correct European coastline bordering the Mediterranean. Africa is shown in a remarkable southward extension, in line with contemporary Portuguese discoveries.

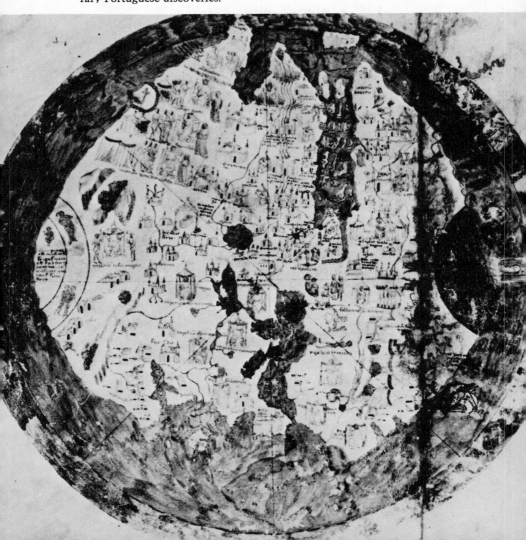

Plethon came to Florence as a delegate to the Ecumenical Council, meeting for two years (1439–40) in the cathedral, in the entourage of the emperor of Byzantium, John VIII Paleologus — one of scores of delegates from all parts of the known world. A Florentine painter who saw this medley of outlandish visitors to the city, Benozzo Gozzoli, has captured the exotic spectacle in a dramatic mural covering all four walls of a narrow, high-ceilinged chapel in the Medici palace. (Giving it a religious title, in compliance with tradition, he called this on-the-scene observation of foreign delegates and Florentine worthies *The Visit of the Magi.*) For our Florentine geographers, the council proved a goldmine of geographic lore, a microcosmic panorama of foreign lands, the *orbis terrarum* come to pay a visit to their native city.

Plethon pulled his full weight in the council's theological sessions. These were by no means mere matters of routine: the pope, Eugenius IV, had convened the council for no less ambitious a purpose than the reuniting of the different churches of Christendom. Amazingly, it achieved its goal — on paper, at any rate.* What prevented this memorable agreement from being carried into effect was, first of all, a storm of indignation on the emperor's return to Constantinople and, ultimately, the fact that the Turks, in their relentless advance, conquered Constantinople in 1453, thereby dislodging Eastern Christianity from its traditional capital.

Yet these arduous council negotiations by no means exhausted the energies of Gemistos Plethon. Though already in his eighties, the old Greek was a fantastically vigorous man. Between proceedings he found the time to hold a series of informal seminars — symposia, really — about Plato's philosophy, in which he found the Italian humanist intelligentsia remarkably uninformed. An outstanding Greek scholar himself, he strove to impart to the Florentine elite a more lucid idea of the ancient Greek legacy. The revival of Plato in the Renaissance goes back to his efforts.

Plethon had still more energy to spare. Like Toscanelli ardently interested in geographic problems, he too went around the council questioning delegates from little-known countries, such as the members of the Russian delegation, about conditions in their na-

---

* The official inscription in which the reunification of the Roman Catholic with the Eastern, or Greek Orthodox, Church is solemnly proclaimed can still be seen in the cathedral of Florence.

Benozzo Gozzoli's VISIT OF THE MAGI (1459), painted on the four walls of the private chapel of the Medici Palace in Florence, in all likelihood shows participants of the Council of Florence, including the emperor of Byzantium, John VIII Paleologus, and members of the Medici family crossing a characteristic Tuscan scenery.

tive lands. Perhaps it was on such an occasion that he first ran into Toscanelli and his friends. He became involved in their problems, engaged them in extensive debates, had Toscanelli show him an unusual map of the northern parts of the earth, let them persuade him to revise some of his own geographic ideas on the authority of Ptolemy — and, above all, taught his new Florentine friends the fundamentals of another great geographer of Hellenistic times, Strabo, who had been little more than a name to the Western world till then. Plethon's influence was so powerful that both the systematic translation of Plato's dialogues and the translation of Strabo's voluminous *Geographika* into Latin were sparked by his persuasive Hellenic fervor.

* * *

For people as urgently concerned about recharting the earth as our Florentines, Strabo was not just another name, hallowed by Greek tradition. How he saw the globe was of the keenest interest.

His vision was brilliant in its conceptual élan. It had been Strabo who had suggested there might be "two inhabited worlds, or even more" inside the Ocean. Could it be that the Ocean sphere was open to navigation as well as to human habitation, so that perchance one might use it as a waterway, sailing from one land formation to the next? If Ptolemy implied something to this effect, the Portuguese expeditions seemed to provide confirmation on both points, at least for the area closest to Western Europe and North Africa and its off-shore islands. Strabo, however, as Plethon could point out, had a definite theory on this whole open problem (epitomizing the best in Greek thought, as Strabo acknowledged by citing his forerunner, Eratosthenes): *"The habitable world,"* Strabo wrote, *"forms a complete circle, itself meeting itself; so that, if the immensity of the Atlantic Ocean did not prevent, we could sail from Iberia [Portugal and Spain] to India [the Far East] along one and the same parallel over the remainder of the circle."*

In other words, if Strabo was right, this would mean that one might reach the spice islands of the Far East by an all-water route, sailing west, using the Ocean as a link between the two far ends of the known world. But even that would leave one more problem: Weren't these islands located in the Indian Ocean, which virtually every map and every authority ever since ancient times described as "land-locked" — an inland sea, blocked toward the east and therewith toward the Ocean by a solid land complex, known since Greek times as the "Golden Chersonese"? Even Ptolemy said so, and quite explicitly, for that matter. If the authorities were correct, reaching the Far Eastern islands by unlocking the Ocean for the use of ships would once more be out of the question.

Strabo denied it. When the discussion got around to this final

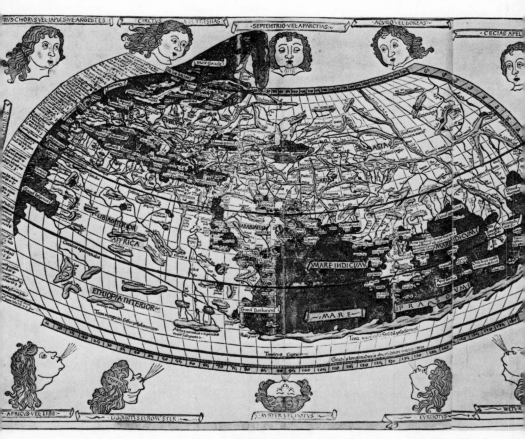

Map from Ptolemy's edition of 1482, done in the Ptolemaic manner. The island world of the Indian Ocean, though already far more diversified than in Bianco's (page 30), is still shown as blocked off toward the east by a solid land mass.

and crucial detail, Plethon may have been compelled to admit that his Strabo was actually a great deal less specific in his Far Eastern geography than Ptolemy, who had considerably more information at his fingertips, writing as he did about a hundred and fifty years later. Nevertheless, Strabo stated as solid fact that the Asian continent was washed by the Ocean on both its eastern and southern shores, and that those Asian islands he knew of were jutting out into the same Ocean Sea. Even if Strabo was in some ways far less specific, his concept was obviously highly attractive, precisely because it conveyed the kind of sweeping global vision, the elementary picture of the earth's surface, that the cautiously empirical Ptolemy did not dare to suggest.

On the face of it, here was one ancient authority's word against that of another. If one believed Ptolemy, the bold idea of using the outer element as a kind of back door to the spice islands would merely confront the navigators with another formidable land mass at the end of their voyage. True, even that obstacle need not be insurmountable; the crews could make the last stretch of their trip overland. But the boldness of Strabo's global vision — and Strabo's authority — would certainly have been impaired if the spice-bearing Far East was, so to speak, turning its back on the Ocean sphere.

We do not know who first thought of checking Strabo's idea against the direct evidence. Maybe the thought was obvious to the Florentines. Quite a few of their Italian compatriots had, after all, traveled to the Far East, bringing back their firsthand data. Marco Polo, for one, had described his voyage from Japan to Indochina and from there into the Indonesian island world as far as Sumatra (and proceeding westward from there, across the Indian Ocean to India itself). The whole glittering arc of islands fringing the East Asian shore, swinging from Japan toward the west, Polo clearly pictured as one continuous chain. Evidently, then, there was no land block between the spice islands and the great Ocean Sea in his Asia.

But just in case there should be any shadow of a doubt, Marco Polo explicitly stated: "And, when I say that this sea [that is, in which Japan is located] is called the China Sea, *I should explain that it is really the Ocean.* But, as we say 'the sea of England' or 'the sea of Rochelle' [the Aegean Sea], so in these parts they speak of 'the China Sea' or the 'Indian Sea' and so forth. *But all these names really apply to the Ocean.*"

Polo could hardly have been more specific. Yet what about Niccolò de'Conti's more recent, in fact contemporary, eyewitness account? Conti had not traveled as far east as his famous compatriot. The easternmost point he reached seems to have been Java (and points roughly on the same longitude on the mainland of Indochina). Still, he had sailed both the Indian Ocean and the China Sea and had formed some idea about their topography.

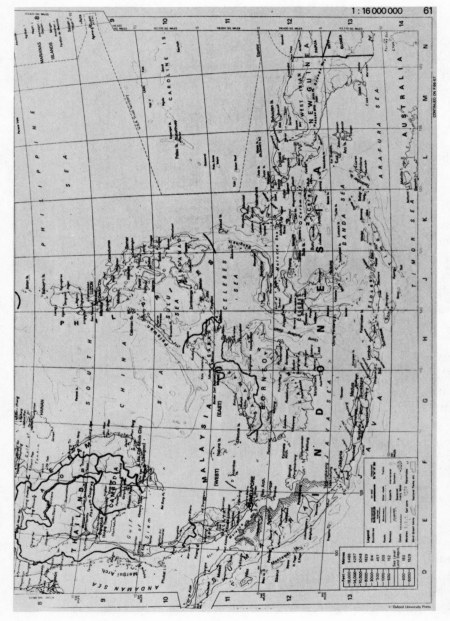

CONTINUED ON P.59-67

The Indonesian island world (shown on a modern map), as Marco Polo and Niccolò Conti saw it — i.e., open toward the South China Sea and the "Eastern" Ocean.

Talking about "Sandai" and "Bandan" — the two islands far-thest to the east of which he had knowledge — Conti briefly raised his eyes toward the unknown world beyond: *The sea is not navigable beyond these islands,* he said (in Poggio Braccio-lini's published account), "and the stormy atmosphere keeps navi-gators at a distance." Putting it differently, beyond these points he perceived the traditional features of the legendary Ocean Sea. Of Java and Sumatra, Poggio Bracciolini's notes observe that "there are two islands toward the extreme confines of the world . . . These islands lay *in his* [Conti's] *route to the ocean.*" Obviously, Conti correctly identified the China Sea — parts of which he had to cross to reach Indochina — with the "Ocean," recognizing that it merged with the "outer sea" (or, as we would now say, with the Pacific).

Conti, in short, may not have been as explicit as Polo when it came to this critical point. Maybe he was more explicit in answer-ing the direct questioning by the Toscanelli circle than he sounds in Poggio Bracciolini's account. What mattered was that Conti, too, knew of no land barrier between the Indian Ocean and the great Ocean Sea — and therewith, in effect, bore out Polo's un-equivocal testimony with his recent observations. Taken together, the two Italian travelers supplied the direct proof to support Strabo's theoretical concept: the spice islands seemed to loom within reach of a sea voyage, no matter how daring or how long. All that was missing was the conclusion that one could actually sail the length and breadth of the unknown sea.

* * *

Trying to reconstruct the thought processes of the Florentines and their dynamic Byzantine friend by glancing at the sources from which they must have pieced together their new map of the world, we see how at last all the elements fitted into a strikingly con-sistent picture. They had no way of knowing by the early fifteenth century that another gigantic continent interposed itself on the route from Iberia to India — another "world," indeed, a New World, as lucid thinkers since Plato, Eratosthenes, and Strabo had for such a long time suspected. Nor could they know that this continent actually divided the Ocean in two. What further hampered their thinking was that they followed an age-old tradition that underestimated the globe's circumference by a substantial extent. Future discoveries — the work of sailors, explorers, cartographers, observers of the overseas flora and fauna — would have to correct the picture and fill in many a critical gap.

Nevertheless, the conceptual picture, the globe in its outlines, was clearly emerging. Through straight, clear-minded thought, involving the ingenious use of the best ideas of earlier ages and the discriminating use of recent firsthand evidence, they were, in effect, unlocking the full scope of the earth. With the authentic methods of rational science they were performing a major creative feat, worthy of the artistic achievements of their time and city — even beyond the incalculable practical results. Dilettantes, scientists, humanist scholars, importers of spices, or whatever they were, in their protracted (and presumably quite informal) discussions they were making a major contribution to the scientific perception of the earth, reviving it from its twelve-hundred-year sleep, placing it on a modern footing.

The Southern Hemisphere was recognized as habitable and, in fact, inhabited. Ptolemy had suggested such a possibility; the direct observations of Polo and Conti about the Far Eastern island world — for the most part located south of the equator — established it beyond doubt. (Even in Polo's time, Dante had still spoken of the lower half of the earth as *"mondo sanza gente,"* an uninhabited world.)

The Portuguese, who did not cross the equator till 1473 or 1474 (and even then only by 1 or 2 degrees), seem to have played no major role in this particular contribution to knowledge. But it was Portugal who supplied the practical evidence for the most crucial feature of the new theory. By proving the Ocean was navigable, at least on its easternmost rim, the voyages of her caravels implicitly corroborated Strabo's whole global concept: if the rim of the Ocean nearest to the western shores of Europe could be sailed (and contained habitable islands, like those settled by the Portuguese), the conclusion that the whole Ocean had to be open for navigation was inevitable. This was as plausible for the fifteenth century as is the inference of twentieth-century scientists, from space flights conducted at a comparatively low orbit, that "space" is accessible to human penetration, in principle, on an unlimited scale and is likely to contain familiar forms of life.

One further consideration must have suggested itself in support of the fifteenth-century hypothesis. If the spice islands were indeed part of the great overall Sea, island-hopping expeditions between those land formations jutting out into the Ocean from the Asian end would be as feasible as those the Portuguese were conducting near the western shores of Europe and Africa. In short, inferences drawn from several observations suggested that the Ocean could be sailed from either end.

Once you visualized the Sea in this way as embedded between the two ends of the known world, its entrance on both sides studded with island groups and inviting the sailor, was it not reasonable to assume that you might encounter further islands, possibly even larger land formations, conceivably entire continents, as you kept heading farther out? Over thirty years later, when Toscanelli recalled the thinking of these discussions, he listed a number of such formations in his letter (using the traditional names common in Medieval myths), and so marked them on his chart. Amazingly, if one takes the trouble to reconstruct his lost chart, which is not too difficult to do from the data contained in his letter, one finds that several of the major land formations Toscanelli envisioned overlap with portions of the actual territory of the New World. Another two years later — though as yet a full seventeen years before Columbus returned from his first voyage to the New World — a Florentine by the name of Lorenzo Buonincontri announced to an audience of fellow citizens interested in the new geographic theories that the existence of a "fourth continent" out in the Ocean was by now considered a certainty. Scientific reasoning, based on logical inferences from the available facts, had projected the New World into the unexplored Ocean sphere years before its actual discovery.

*　*　*

Thirty years is a long time, even in an old man's life, when the years seem to be speeding by faster and faster. The excitement of those early debates had evaporated long ago. The companions had dispersed or were dead. One generation had passed away; another had come. Gone was Cosimo Medici, under whose cautious but generous rule the new geographic ideas had ripened. It was Cosimo who had patronized and encouraged the new style in building and the new vision in art, who had shared the fervent interests of the humanists and had invited the Ecumenical Council to Florence, including Gemistos Plethon, the fiery old scholar from Byzantium. Prince Henry the Navigator had died a few years before Cosimo, and for a while the Portuguese had slowed down in their expeditions. Now, under Afonso V, they were revitalizing their ambitious venture.

> I have already spoken with you [Toscanelli wrote to the canon of the Lisbon cathedral] respecting a shorter way to the places of spices than that which you take by Guinea [Africa], by means of maritime navigation. The Most Serene King now seeks from me some statement, or rather a demonstration to the eye, by which the slightly learned may take in and understand that way ... I, therefore, send to His Majesty a chart made by my own hands, on which are delineated your [that is, the Portuguese] coasts and islands, whence you must begin to make your journey always westward, and the places at which you should arrive at those most fertile places full of all sorts of spices and jewels. *You must not be surprised if I call the places where the spices are West, when they usually call them East, because for those proceeding by navigation over the lower side of the earth these areas will always be found to the West. But if by land and on the upper side, they will always be found to the East ...*

Perhaps this was not the sort of exact terminology a modern scientist would employ. Neither was it a very explicit statement. Yet for a contemporary reader — for Fernão Martins, for Afonso V or Columbus — Toscanelli evoked a picture that must have been as intelligible as it was striking and at the same time authoritative: You can reach the spice islands by sailing west across the Ocean as well as by going eastward via the customary overland routes. Overland you would be traveling within the Northern Hemisphere; on the sea route you would have to cross the equator (traveling on the under side of the earth).

Toscanelli must have known from Polo and Conti, if through no other source, that the bulk of the Indonesian island world is in fact situated in the Southern Hemisphere. On this point he could correct Strabo, who envisioned the Ocean crossing to be conducted "along one and the same parallel." In every other respect this was the Strabonic hypothesis hardened into certainty by solid empirical thought.*

The Portuguese, notwithstanding this letter, persisted in their traditional course around Africa. But when they had finally rounded the Cape of Good Hope and actually reached India, someone by the name of Pietro Vaglienti (who was also active as an importer of spices and, as a younger man, had known Toscanelli) attributed even that feat to the vigor of Toscanelli's geographic concepts. Quite correctly, too, because whether you were sailing west across the Ocean or around Africa, you were still operating within the same conceptual framework in viewing the globe as an all-accessible entity, with the Ocean a potential link between the known lands, a waterway between the two ends of the three-continental land mass.

Even Columbus, who possessed no more than a smattering of knowledge of geography, had a much broader debt to Toscanelli's circle than merely the receipt of a copy of that famous letter. His son Ferdinand, a highly cultured man who carefully compiled the data of his famous father's life, acknowledged that Columbus' chief inspiration had been Strabo's concept of the globe, which was first integrated into Western thought through Gemistos Plethon's discussions with his Florentine friends. Columbus' "enterprise of the Indies" (as much as any other major voyage of the age of discoveries) ultimately rested on Strabo's premise that "the habitable world forms a complete circle, itself meeting itself" or that the whole globe is inhabitable and accessible to navigation — a premise verified by the Florentine geographers through their careful use of fresh evidence.

All the great expeditions of the age of discoveries proceeded within the theoretical framework that had been established by a handful of Florentine humanists — that ingenious picture puzzle of the globe compiled from ancient and contemporary sources, with a conceptual flair characteristic of the Renaissance mind, including a first hazy outline of the American continent.

* Conceivably, the phrase "over the lower side of the earth" may even have had a more elementary meaning. In the traditional conception, the habitable land mass formed a kind of "cap" covering the top of the globe, and the Ocean enveloped its lower portion. Toscanelli may have had this simple picture in mind, which was, in fact, quite typical of contemporary world maps.

TWO

Ancient Roots

A CIVILIZATION like our own, unique in its submission to the rule of science and technology, is naturally curious to know how science and its power over our culture may have come about. Indeed, the history of science is a discipline created in modern times. Any self-respecting bookstore these days is likely to display a well-stocked shelf of paperbacks in the history of science. Some are general histories; some trace the evolution of special fields — mathematics, physics, chemistry, and so on. Many of them make excellent reading. Most of them can deepen our sense of history, opening up the dimension of the past underneath an otherwise somewhat casually accepted present. All of them, however, seem to suffer from one particular flaw: they fail to relate the evolution of science to the overall historical process. The histories tend to present science as a hermetically secluded, gloriously isolated activity of the mind, with little or no relation to such grubby experiences of the common mass of humanity as wars, revolutions, epidemics, floods, or whatever other vicissitudes are known to intrude on the quiet pursuit of a particular line of thought.

* * *

A series of disasters without parallel had interrupted all meaningful continuity between the ancient and the Medieval world, obliterating creative science in the process. A singularly lucky combination of circumstances rekindled scientific thought around the twelfth century, and thereby gave birth to a new and continuous phase that has by no means yet reached its peak. So fortunate, in fact, was that historical constellation that it sparked the most spectacular scientific evolution history has ever seen so far.

The eclipse of science in the early Middle Ages was a direct result of the eclipse of virtually all cultural life, brought on by the breakdown of Roman civilization in the West. The revival of science around the twelfth century was, in turn, associated with a general cultural revival, which led in direct ascent to the great cultural flowering of the Renaissance. Far from occurring on some splendidly isolated level, the major ups and downs in the history of science seem to reflect the principal movements in the broad course of cultural history. So do some of its major directions and insights. How people saw their cosmos, how they perceived their earth, was intimately conditioned by their cultural situation — which was in itself determined by the patterns of causality that rule the broad flow of historical events.

The exploration of nature obviously can be prompted by any number of motives, many of them perfectly practical. But the almost total eclipse of science in the early Middle Ages suggests that the most consistent and fertile scientific endeavors are nourished by what one might call a "secular" attitude, a curiosity about nature rooted in a secular cultural climate — one in which the immediate sense perception plays a central and fully recognized role. Clearly, the ancient world had attained just such a climate in the pantheistic, this-worldly cultures of Greece, of Hellenism, and of Rome. And it was just this sturdy acceptance of nature, this cheerful delight in its gifts, that was abandoned during the early Middle Ages and to which Europe was at long last returning — gradually and with difficulty and often with grave reluctance — during the long process of secularization between the twelfth and the fifteenth centuries.

Science seems to grow most luxuriantly in cultures with a positive attitude toward the world of the senses; it appears to wither in cultures with an emphatically spiritual, otherworldly bent. The evolution of science bears, therefore, a strong kinship to the more sense-oriented phases in the history of literature and art. And it proceeds under a cloud of inherent antagonism, if not toward religion, at least toward cultures with strong transcendental tendencies, usually sanctioned and rationalized by religious beliefs. Science had started in the earliest civilizations, in the Nile Valley and in Mesopotamia, and is therefore as old as history itself. In fact, since our knowledge of prehistoric times is still scanty in the extreme, we must admit of the possibility that respectable scientific achievements should be credited to prehistoric cultures. The recent decoding of the astronomical calculations that lay behind the arrangements of the great boulders at Stonehenge, in the south of England, challenges our modern arrogance, which makes us assume that "primitive" people were incapable of astonishingly sophisticated natural observations. If the history of science could teach us any general lesson at all, it would be one of thorough respect for the intellectual capacities of our ancestors, no matter how far back in the past.

It is a mark of modern ignorance to think that we have become progressively smarter. As far as sheer mental vigor is concerned (including the capacity for applying a disciplined intellect to a given problem), history tells us in no uncertain terms that creative intelligence was always present as a human potential, even in civilizations that we choose to think of as primitive. It is the body of knowledge that has changed and expanded (along with the methods for classification and basic approach), not the mind or its powers. Who is to say whether the task of tackling a problem without the benefit of a well-developed body of methods and information may not have required far greater intellectual vigor and originality than is needed for proceeding from problem to problem within the safely established disciplines? Prehistoric, early historic, as well as Medieval science have faced such a task.

* * *

Most early civilizations lived under some sort of theocratic system, as a matter of fact. But so formidable were the problems of a settled existence — a new and unaccustomed experience after the long span of nomadic life — that people demanded factual answers, no matter how spiritual the orientation of the prevailing cultures may have been. Agriculture presented such a new type of social experience; even more so, the life of the new urban communities arising in the valley between the Euphrates and Tigris, in Egypt, and along the banks of the Indus in the Punjab. The new

Stonehenge, in southern England, probably built as a temple devoted to sun worship, served at the same time as an amazingly accurate instrument for computing eclipses. The rays of the rising sun show directly on the heel stone during the vernal equinox on this prehistoric forerunner of Toscanelli's gnomon.

experiences set off a spurt of inventions, the "neolithic revolution," a phase almost as rapid and intense as the Industrial Revolution of our modern world. They stimulated the mind and raised questions about the rhythm of the seasons, the flooding of the great rivers, the natural relations of the forces of heaven and earth; questions about the construction and equilibrium of monumental structures; about foreign countries and their products; about road-building and trade. These new experiences stimulated technology and astronomy, statics and mechanics, mathematics, geography, botany, zoology, navigation, and medicine.

On the empirical level of fact-gathering and straight observation, the sciences flourished in the vibrant climate of the earliest cultures. The rise of civilization in the ancient East was extremely conducive to basic scientific thought and technology, even though the religious explanation of the workings of the universe tended to block the emergence of a full-fledged natural philosophy.

At last, scientific thought reached a climax with the brilliant intellectual freedom of Greece. The Greek mind was remarkably free of transcendental restrictions, which would reserve hazy areas to the mysterious action of the gods. The Greeks lived in the here and now, in a lucid world open to human action and human perception; their gods on Olympus were part of the Greek landscape and Greek life, an intimate part of their cultural scenery. The feeling of powerlessness before the elementary forces of nature was lifted. The observation of nature became imbued with a pantheistic sense of poetry; science became a facet of philosophical thought.

The leap from the mythological cultures of the ancient East to the world-minded Greek culture raised science from a level of scattered empirical observations to the order of a consistent natural philosophy, with cosmology its main object. The Greek mind was the first to replace a mythical, or religious, view of the cosmos

The Zeus (or Poseidon) of Artemision, from the fifth century B.C., embodies the Greek sense of self-confidence and freedom in confronting nature.

Amphora with satyrs treading grapes, from circa 530 B.C., expresses some of the lighthearted approach to nature that distinguished the Greeks.

with rational interpretations; the first therewith to expand the scope of science to the dimensions of the universe. In this sense, the popular idea that science began with the Greeks is justified. Even if respectable though scattered beginnings had preceded the Greeks by some three thousand years, we may think of Greek science as the first systematic and inclusive attempt to explain the entire natural cosmos.

*   *   *

What freed the Greeks for this staunchly rational approach to the natural world? What made them look at the world in cheerful reliance on their senses, in easy companionship with the gods — no longer intimidated by those fears that had caused earlier cultures to shrink from a rational explanation of the cosmic forces?

Greek civilization arose from a combination of two cultural legacies, both marked by powerful elements of secular freedom. Descendants of Indo-European tribes that had broken into the Balkan Peninsula and the island world of the Aegean, the Greeks inherited from their nomadic forebears a natural vigor and independence of mind. Their ancestors had settled in the craggy mountain valleys of Greece, across the myriad of islands of the Aegean or along the Asiatic shore, preserving basic tribal freedoms in their communities, whose very compactness encouraged the rise of democracy in the end. A streak of rugged personal independence perpetuated itself from roaming nomadic days through the establishment of the *polis*, to the eventual full participation of the citizen in the running of his city-state. A combination of ethnic and geographic features had saved the Greeks from those monolithic government structures that had stamped out individual freedom in the earlier cultures.

There was, too, a tradition of earthy enjoyment of life, of commercial prosperity and esthetic refinement awaiting the newcomers in the ancient Aegean culture, with its center on the island of Crete — a culture that flourished all over the area the invaders had conquered. Greek culture, the product of this happy blending, combined the sophisticated lifestyle, the pleasure-loving, easygoing ways of the Aegean merchant or sailor, with the rugged independence of the tribes as preserved in the city-state.

Greek culture was the creation of free tribal men (who managed to assert their freedom against a succession of native tyrants, even against the onslaught of the colossal Persian Empire; and who had the good fortune in the bargain to inherit the gusto of the freest among the early cultures). The Greeks looked at a world that was theirs through conquest and seemed beautiful and pleasurable by cultural inheritance. The world of nature seemed an open domain for the enjoyment of the senses and the comprehending penetration of the mind.

From its earliest manifestations the Greek mind had turned to natural philosophy: Greek cosmological speculation began with the pre-Socratic philosophers. The first stirrings of Greek philosophical thought were identical with the beginnings of Greek science. That was how graciously the world of nature beckoned to the Greek mind.

Thales of Miletus, in the early sixth century B.C., had seen the cosmos as a watery universe, enveloping the disk-shaped earth on all sides. To Thales, the sun, moon, and stars were vaporous bodies, sailing in a state of incandescence across the liquid firmament till they gently dropped into the cosmic sea, where they floated back to rise once more in the east. Obviously, Thales had magnified the geographic concept of an all-encircling Ocean to cosmic proportions. The ancient view of the earth, with its characteristic predominance of the Ocean Sea, had been expanded into a vision of the universe. Astronomy and geography were still bound up in one common embryonic vision in the beginnings of cosmological thought, even as science and philosophy in the beginning were one.

If Thales' poetic vision strikes us as naïve (and is in its substance still virtually identical with the universe the mythological "cosmogonies" of the ancient Near East had envisioned), it is a first scientific cosmology nevertheless. As he saw it, cosmos and earth had been formed by a natural process, no longer through an act of the gods.

The cosmic image had again been expanded and diversified by Anaximander, another Ionian philosopher (that is, one who, like Thales, came from the eastern shore of the Aegean Sea), whom history describes as Thales' successor. Anaximander had visualized the cosmos in a state of continuous evolution; had conceived of an original substance he was sophisticated enough to call "the Undefined"; had thought that the world was created through a series of explosions produced by the pressure of fire on water and "mist" (or, as we might say, oxygen); and had pictured the process of evolution in the form of fiery vortices wheeling around in cosmic space. The human species, Anaximander thought, had evolved from more primitive animal forms, of which the first were amphibious — what Anaximander called "the fish."

Water as a primordial substance seems to have enveloped early Greek thought — even as it loomed over the rise of geography; or as it bathes the opening lines of the ancient religious myths, including the Bible, where "the Spirit of God moved upon the face of the waters." What is so startling about Anaximander's thought is that within the space of two generations he had advanced Greek cosmology by the sheer force of speculation and intuition (plus an occasional dash of empirical reasoning) from the early mythical notions to a point that, in an amorphous sort of way, is surprisingly close to our modern concepts.

*   *   *

It might seem that the human race had been launched on an unbroken voyage of growing insights into the natural cosmos. But that was not to be the case.

Greek science — and the Greek picture of the cosmos — continued to flourish, along with the growth of Greek cultural life. But when Greek culture began to wither away at its roots (first from the murderous conflicts among the city-states; in the end through the conquest of an overpowering neighbor, Philip of Macedonia, and his son, Alexander the Great), the broad flow of human events once more asserted itself, washing away with its swift currents the tenuous plant that had flowered into Greek drama, Greek philosophy, architecture, and art. Scientific thought, like all outstanding cultural creativity, seems to flourish only in a particularly nourishing soil. The forces that came to toss about the late ancient world proved increasingly hostile to the climate of contemplative leisure that breeds the undisturbed growth of original thought.

Yet the end did not come with brutal abruptness. A tradition of science inspired by Greek thought managed to maintain itself against the upheavals of the following centuries, until it was at last silenced (in the Western world, at any rate) by the collapse of Rome. Aristotle, Alexander's teacher, saved the legacy of Greek speculative thought and implanted it in the cosmopolitan civilization, blended from Greek and Oriental elements, that sprang up all over the empire Alexander had built with his sword. In fact, largely because of its cosmopolitan nature, the culture of Hellenism, which was eventually supported by the political institutions of the Roman Peace, fostered great scientific and technological vitality. Though Hellenistic science hardly sustained the creative vigor that had marked early Greek thought, it added a sense of preservation, orderly classification and critical sifting, and, above all, a feeling for the concrete detail, which contributed greatly to the systematic evolution of scientific disciplines.

An atmosphere of sober empiricism distinguishes Hellenistic science, which proved invaluable for all future scientific growth. Some Hellenistic scientists, like Eratosthenes, Strabo, and Ptolemy (the last especially in his astronomical thought), still showed a spark of that original speculative élan with which Greek science had started, just as Aristotle, who stood at the divide between Greek and Hellenistic culture, had managed to channel the power of Greek philosophical thought into a logical system of scientific classification and method. His system came to exercise an enormous influence over the following two thousand years.

The Hellenistic — and, with it, a final spark of the original Greek — tradition survived in the cultures of Byzantium and Islam. It was mainly from these sources that the West, after a long night of low cultural vitality, managed to rekindle the flame of trenchant cosmological speculation and inspired natural philosophy.

* * *

Not long before the Greek mind began to falter under the blows of historic adversity, it had attained its most sublime vision. The two centuries preceding the end of Greek independence saw the consummation of Greek culture, the "golden age" of Greece. Side by side with the statues on the Acropolis, the plays of the Greek tragedians, and the comedies of Aristophanes, the genius of Greece gave birth to a cosmic vision that found its most brilliant scientific expression in the philosophy of Pythagoras and its most profound metaphysical expression in Plato's philosophy. If Plato's insights were of such power and depth that, as Alfred North Whitehead has said, all subsequent Western philosophy "consists of a series of footnotes to Plato," one might add — trying with similar exaggeration to capture the essence of an intellectual feat — that all later concepts of the natural universe were elaborations or modifications of Pythagoras' original insight. The great cosmic visions seem to present themselves only at the cultural heights of humanity.

* * *

Pythagoras came from Ionia — from Samos, an island in the immediate vicinity of the Aegean east coast, just across the gulf from Miletus, where Thales and Anaximander had lived (and had completed their theories by the time Pythagoras grew up). The first rumblings of the Persian wars, Ionia's conquest by the Persian colossus, drove him into exile to Croton in southern Italy, one of the colonies founded by the Greek city-states in that part of the world.

Perhaps it was the experience of exile, the uprooting and forcible adjustment to a new set of cultural values, that made Pythagoras depart from the Ionian tradition of natural philosophy. Perhaps his thought reflects the intense self-reliance that the exile is often compelled to adopt, and that may cause him to seek a new home in a spiritual communion with the universe.

Whatever his motivations, such elements characterize Pythagoras' mind. His thought is religious and spiritual. But instead of returning to the mythologies of the past, he infused scientific speculation with a new spiritual conception of the natural universe. Pythagoras gave the cosmos of the Ionians a spiritual dimension and a metaphysical depth.

The Ionians had conceived of nature as a completely self-motivated entity. The workings of the universe occurred as mere extensions of the primordial chaos, automatic functions of its basic elements. Matter possessed its own evolutionary qualities. "Order" and "law" were mere concepts superimposed by the human mind on the autonomous processes of nature. Nature herself knew of no laws. It was Pythagoras who introduced the vision of an intrinsic natural order. No wonder Pythagoras (and Plato, who adopted his vision) came to command the most powerful influence over the natural philosophy of the Middle Ages and Renaissance — and, as a result, over the basic conceptual framework of modern science: they are the fathers of the idea of natural law. Pythagoras' mind created order out of the primeval chaos.

* * *

One way to describe the Pythagorean insights would be to say that the Greeks before him had conceived of the universe chiefly in terms of matter. Pythagoras (and his school) perceived the essence of form.

"Form" — the fact that matter presents itself under definitely structured conditions and develops (or "moves") according to definite patterns or laws — is ultimately a mathematical phenomenon. To put it differently: form in its stationary aspects is a phenomenon that is measurable in geometric terms; and the motion, or evolution, of matter occurs in relationships that can be measured in algebraic terms, which may often, in turn, be translated into the figurative terms of geometry. Yet mathematics is not just a "language," a way of *describing* the proportions of static form or of *describing* the relationships of evolutions or movements. Mathematical relationships tend to show a forever-surprising basic simplicity, as if implying that certain, relatively few fundamental laws, or their variants, underlie the infinite multitude of the observable detail that offers itself to our senses. To discover that the universe is structured, and moves, according to mathematical laws is to experience one of the most profound insights into the basic order of the cosmos.

Perhaps mathematics is indeed merely a "category of our thinking," as Kant has said of space and time. But every phenomenon in nature presents itself in the curious language of mathematics. By the same token, almost any venture in the practical mastery of the forces of nature — any engineering feat, to name an example — has to be translated into the language of mathematics before it can be tackled successfully. Mathematics is the vocabulary by which the human mind apprehends the intrinsic order of nature, communicates with it, and masters its laws. If it is not the essence of that order itself, it is at least one of its most fundamental functions or aspects.

It would be too much to claim that Pythagoras and his followers correctly apprehended all basic mathematical relationships; but they perceived their existence and went surprisingly far in anticipating their scope. The Pythagoreans discovered that musical notes differ in height according to basic mathematical relationships (which Pythagoras measured in the tensions of vibrating strings and the varying lengths of reed pipes); and that music therefore is another expression of that same inherent order that speaks to us through the mathematical vocabulary. They realized that mathematics, like music, is ultimately an expression of harmony. Their having perceived that particular harmony, in whatever guise or on whatever level, links their insights with the innermost quality of Greek art and the Greek crafts — the ideal of harmony realized in a Greek statue, the elemental harmony and symmetry of a Greek vase. (Plato, it happens, was keenly aware of these associations.)

It was harmony that Greek philosophers like Plato saw as the ultimate aim of the soul — a dynamic harmony between contending forces or conflicting tensions, not some lifeless condition of "peace"; harmony as a facet of beauty, which Plato described as the most profound human insight into the essence of the Divine. In that sense, Pythagorean mathematics, Greek thought, and Greek art are really concerned with an identical substance; it is at the very core of the Greek way of life.

The Pythagoreans knew that mathematics underlies any effective practical approach to the natural world. They also anticipated the basic laws of Euclidian geometry. If they were wrong in assuming that the planets move in "perfect circles" (an assumption that was later included in Aristotle's astronomy and accepted until Kepler's time), the fact that the planets actually move in elliptical orbits merely reveals a more complicated but certainly no less cogent mathematical law. An even more astonishing concept of the Pythagoreans was that the earth (along with the planets, as well as the sun) revolved around some great cosmic fire located at the center of the universe. They therefore conceived of the earth as a planet, and at the same time perceived its spherical shape.

A mathematically structured cosmos, evolving according to basically simple laws of inner order and harmony; a cosmic projection of the Greek sense of the underlying dynamic harmony of all being — that was, in essence, Pythagoras' vision of the world. His conception has influenced our thinking to this day.

* * *

It is strange how closely our picture of the cosmos depends on our cultural environment and is at all times shaped by basic historic forces. In Western scientific thought, the idea of natural law was not reborn before the School of Chartres in the twelfth century. The concept of a mathematically structured cosmos was not revived, as an explicit insight, until the mathematical speculations of the fourteenth century; not perceived in its full esthetic meaning until the fifteenth-century Renaissance; not grasped in its complete scientific grandeur until the seventeenth century, with the "universal mathematics," the *mathesis universalis,* of René Descartes. During the intervening two thousand–odd years, the cosmic image was at first refined — then sorely mutilated and disfigured, a victim of historical turbulence.

Aristotle, at the dawn of Hellenistic civilization, had compressed the Pythagorean cosmos into a solidly rational, enormously complex (and rather rigid and mechanical) framework.

Ptolemy, by the end of that age, had changed it along more flexible, more sophisticated, though in fact even more complex lines. But once the floods of barbaric invasions had washed over the ancient world, what remained to the early Middle Ages of the Greek cosmos was a sapless and puerile picture, such as a tent spanned over a rectangle — the tent presumably being God's heaven, the rectangle (or "God's footstool") being what was left of the image of the earth.

The breakdown of ancient civilization had played brutal havoc with the cosmic imagination of Western man.

* * *

The immediate reason why the inspired cosmic vision of the Greeks had shriveled to this childish picture was that the observation of nature was deliberately stunted in the Western world, under the impact of the collapse of Rome. To those who lived through this catastrophe it seemed that the utter breakdown of civilization had come, the ruin of everything humanity had ever tried to create over thousands of years, a verdict from a wrathful heaven. The West reacted with a radical readjustment of the mind.

The earth, this poor vale of sorrows in which the collapse of civilization had taken place, no longer seemed a worthy object of intellectual scrutiny. In a great wave of abysmal disillusionment, the mind began to negate the experience of the senses. The West was turning its back on sense perception, on the earth, on the observation of the natural universe, on Greek and Hellenistic science, on everything that seemed tainted — however remotely — with the traumatic memory of that shattering collapse; and on the whole sense-oriented, life-loving outlook that seemed to have precipitated the disaster. The end of Roman civilization in the West (all the more crushing because it occurred in a series of long-drawn-out and repeated barbaric invasions, like the outbreaks of a volcano that refuses to subside) would have left a mood of collective despair, an orgy of bleak and impenetrable hopelessness, had it not been for the inspiration of Christianity.

Christian teachings offered a hope for the victims of the catastrophe. The hope — the only hope the gigantic disaster could conceivably leave in its wake — rested in the conviction of the essential unreality of the things of this world, in contrast to the abiding reality of an invisible world that, by its very nature, had to remain immune from the shocks of secular experience. If the original Christian faith had contained certain world-denying elements, a handful of religious philosophers, the Latin Fathers of the Church, reinterpreted basic Christian teachings to suit the new mood of the Western world in order to alleviate the despair of that afflicted community. Where original Christianity held out the hope of an afterlife by stressing the chance of salvation for every follower of Christ who would mold his life to prepare himself for heaven, the new "Western" Christianity asked for the complete denial of the world of the senses, asserting that life in this world is neither of primary significance nor, in a philosophical sense, actually "real."

The strange mentality that emerged from this theological structure looks to us as though all elementary experience had been not only denied but deliberately reversed. Yet it was probably the only ideology by which the West could hope to survive. It proved its vigor and vitality by ruling the Western mind for almost a thousand years.

* * *

Manifestly, there was no room for scientific observation inside this Medieval, transcendental view of the world. Nor was it possible to accept this curiously inverted outlook and at the same time maintain any kind of rational vision of the natural cosmos, which must, after all, be based on some degree of empirical observation. The Medieval cosmos was made up of faith and fantasy, and often profound metaphysical insights. We would call it "unreal" (or "supernatural"), even though to the Middle Ages it was brimful of life and possessed a higher reality than our ordinary daily experience.

The astonishing fact is that the Middle Ages, for close to a thousand years, seriously believed in a "transcendental universe," in which the phenomena of the visible world were merely a poor reflection of the life of the higher spheres. It was this belief that gave Medieval culture its mood and its flavor. It stamped Medieval life with an inwardness, a quiet, spiritual quality, and a sense of stoic acceptance that the modern world seems sorely to lack. The idea that material things are not really important (conceived at a time when the experiences of the "real" world had become almost unbearably harsh) was exuding its quiet charm over the daily life of the Middle Ages. Much of that lovely spirit of inwardness that still speaks to us from the stones of a Medieval church or the colors of a Medieval painting really stems from that insight — born of the tragic collapse of ancient civilization — which was first and most convincingly phrased by St. Augustine, the prime architect of the Medieval mind. Yet this whole poetic outlook and life style spelled doom for any scientific pursuit.

> When . . . the question is asked what we are to believe in regard to religion [St. Augustine wrote], it is not necessary to probe into the nature of things, as was done by those whom the Greeks call *physici;* nor need we be in alarm lest the Christian should be ignorant of the force and number of the elements — the motion and order and eclipses of the heavenly bodies; the form of the heavens; the species and the nature of animals, plants, stones, fountains, rivers, mountains; about chronology and distances; the signs of coming storms; and a thousand other things which those philosophers either have found out or think they have found out . . . It is enough for the Christian to believe that the only cause of all created things . . . whether heavenly or earthly . . . is the goodness of the Creator, the one true God.

With one mighty stroke of his pen, the great saint had banned an almost complete catalogue of the ancient sciences — physics, cosmology and astronomy, zoology, botany, geology, hydrology and hydrography, history, geography, meteorology. As he sent his contemporaries into the adventure of Medieval civilization, he told them to raise up their eyes to heaven and, most explicitly, forget about the things of this earth. In a world ravaged by barbarian invasions the contemplation of the heavenly spheres had to replace the study of nature.

Once the mind had been given this orientation, science could not simply be taken up again, without fundamental changes in the cultural framework. To undertake any observations of nature in a straightforward way involved both a radical change in the accepted view of the world and a complete revolution in mental habits, in the ways a given problem was habitually approached. This is why the revival of natural observation called first of all for a thorough philosophical critique of the dominant transcendental ideology and its methods, a complete revision of the established frame of mind. Even more basically, this is why the resumption of science had to be part of an emphatic reassertion of sense perception, a broad and conscious movement of a return to nature and to the earth.

* * *

In purely chronological terms it might seem as though it took Europe something like eight hundred years to recover from the collapse of Rome — from the early fifth century, when St. Augustine witnessed the first major shocks of the collapse of the ancient world (in the West, at any rate), to the twelfth, when a remarkable upsurge of trade generated the first vigorous stirrings of a new cultural vitality, primarily in France. But such astronomical figures make us overlook the realities of the historical process.

The ebullient economic and cultural life that pervaded the twelfth-century "Renaissance" belonged in essence to a new civilization that had little in common with the defunct world of ancient Rome, beyond the memory of a common heritage, a thin thread of tradition. What looks like a "revival" of ancient culture was a revival only in the most general sense of the term, a revival of energy and vitality, even though the participants in this experience more and more thought of it as an explicit rebirth of antiquity, a literal "Renaissance." Actually, the new prosperity — and the new flowering of culture — was not brought about by exactly the same ethnic groups that had lived under Roman civilization, nor did it flourish over precisely the same geographic area, nor was it motivated by the same attitudes or outlook upon the world. It was a new culture, whose singular originality hides behind that drab label applied, in hindsight, by the humanists of the Renaissance, *medium aevum*, the "Middle Ages" — a term suggesting that this was not a culture in its own right at all, but merely a sorry interlude between two glorious civilizations, ancient and modern.

As we now know, it was Medieval civilization that in almost every respect produced the archetypes and the dynamic momentum for the main forces of the modern world. Dynamic and deeply original itself, Medieval culture represents the gestation period in the rise of the modern West. It was the great caldron in which the broad stream of the ancient legacy was remolded and recast until its major components emerged in their basic modern shape. Yet, at the same time, the Middle Ages developed an original culture of unique creative vitality. Modern science and the new natural cosmos, along with capitalism, modern parliamentary democracy, labor unions, modern social systems and political thought, and the foundations of modern technology are among the products of Medieval culture — together with the great Gothic cathedrals, the insights of Thomas Aquinas' philosophy, or Dante's poetic vision.

Who were the builders of Medieval civilization? They were Germanic tribes, Indo-European nomads or seminomads, whose intrusions into the ancient world had brutally interrupted the flow of events more than once. The builders of Greek civilization had come from the same background (and the same general sites) as had the founders of the caste system and the Hindu religion in India; also, in all likelihood, some of the fiercest tribal invaders of ancient Mesopotamia and Egypt, as well as the "Italic" people, the founders of Rome. Finally, after a great clash with Mongol nomads — a kind of gigantic world war among prehistoric hordes — the Germanic tribes had once more swarmed into the civilized world along the Mediterranean, infiltrating across the Roman borders, peacefully at first, yet in the end washing away the empire to its very foundations.

The turmoil did not subside for centuries. But slowly the nomads had begun to create their own civilization. They used many of the materials they found — the remnants of Roman roads and cities, some of Rome's legal and political institutions, some sparks from the ancient cultural legacy. Essentially, though, the civilization they built was distinctly their own. They imbued it with the characteristic technical ingenuity of the nomad (thereby setting off a vigorous evolution of technology whose impulses swing on into the modern world). They formed a singular new type of social organization — the feudal system — made up of elements they saw in the conquered Roman territories, yet mostly of their own tribal customs; a curious system, at once very primitive and very involved, that was exactly suited to the chaotic situation in which they happened to find themselves.

They learned to live within the cultural ideology of the Christian Church, including the strong transcendental leanings that had been instilled by St. Augustine and the other Latin Fathers. It proved a useful ideology for the frontier-type of life that ruled the early Middle Ages. The world-denying spirit that had been so trenchantly imprinted on the Christian faith by the collapse of Rome helped to sustain a staunch, self-negating morale in the face of physical hardships and the insecurity of elementary living.

Around the turn of the millennium the venture of Medieval civilization had turned out to be a ringing success. A handful of ingenious technological inventions had multiplied the productivity of the land. The rule of a military caste, implicit in the feudal order, had at last brought peace — basic military security from the perennial scourge of nomadic invasions. Agricultural productivity, technical ingenuity, and the relative peace combined to spark vibrant trade and, along with it, the first primitive manifestations of a budding industrial life, which has increasingly absorbed the West ever since.

The great abyss opened up by the collapse of Rome was beginning to close. The area of the new civilization, dynamically reaching toward the European North, Northeast, and East, outstripping the old Roman borders, began to reap the blessings of a burgeoning economic prosperity.

* * *

Anyone open to intellectual stimulation who looked around the twelfth-century scene might have ascribed its ebullience to the merest chance. Lady Fortune (whose dainty fingers the Renaissance was to suspect behind every turn of the wheel of history) might have jumbled historical circumstances like the little colored pieces in the kaleidoscope, with the most fortunate results indeed.

Everything seemed to conspire for a reawakening of science on a new and magnificent scale. Trade and technological experimentation were powering the life of the cities. Counting houses, warehouses, ports, markets and shops, the workshops of craftsmen and the homes of merchants, even the Gothic cathedrals going up within the narrow space of the city walls — all were pervaded by the new opulence; everything breathed the spirit of innovation, the adventure of experiment. Every cargo unloaded from ships (equipped for distant voyages by an array of new features in their design), every outlandish commodity displayed on the market stalls, every new luxury item in the residential houses, spoke of the natural bounty of the earth, the excitement of foreign cultures. The Medieval towns were becoming a part of the wide-open world — though they were still hedged between their ancient walls, still overshadowed by cathedral and castle.

If someone had stood on the city walls and turned his head toward the countryside, he would in effect have looked into an older society. The feudal order still dominated the country, its somber castles towering above the hills, its strongholds firmly planted even inside the city walls. By now the new commercial classes had begun to lash out against the feudal powers, trying to dislodge them from their castles and to wrest the control of the towns from their grip. But amazingly, even the feudal order, merely by its continued presence, contributed to the climate of initiative and experimentation.

It was under the feudal system that the native ingenuity of the Indo-European nomads had been channeled into a dynamic technological evolution. Now, feudal restrictions on trade were challenging the resourcefulness of the merchant class to keener exertions, which proved crucial for the rise of the capitalist economy. Faced with the ever-present feudal dues, tolls, and other obstacles for commercial initiative wherever he turned, the businessman had to work that much harder in order to net a rewarding profit. He frantically tried to expand the scope of his trading, to search for new markets, to mobilize all his inventiveness in fostering an elementary industry. Without the spur of feudal restrictions, the growing prosperity might have resulted in merely another commercial economy, such as the world had often seen before. The challenge of feudalism helped to turn the merchant into a restlessly searching, endlessly resourceful entrepreneur, and trade into early capitalistic enterprise. It was feudalism that had fathered the rise of Western technology and that now unwittingly nurtured its first industrial applications.

In this early and fortunate state, even the otherworldly tradition of the Church yielded major incentives for the rebirth of science. "Faith in the possibility of science," Alfred North Whitehead said, "generated antecedently to the development of modern scientific theory, is an unconscious derivative from medieval theology." Not only a faith in science's potentialities, we might add: the whole intellectual climate in which early Western science ripened was steeped in the traditions of the Medieval Church. If science was to take wing again after its thousand-year sleep, St. Augustine's vision had to give way to a vigorous emphasis on the natural world. But these Augustinian origins were to affect decisively science's whole future shape. The need to match the transcendental universe with a new natural cosmos — which had to be equally universal and equally grounded in philosophical argument — imbued modern science with its universal tendencies, its characteristic ideological aspirations, its inherent logical discipline. Just as the presence of feudalism forced trade into early capitalistic channels, the powerful hold of Medieval Christianity on the European mind forced science to start out as a consistent intellectual system. Without that ever-present challenge, science might have been reborn as no more than a rich but unsystematic body of observations, conducted along pleasurably positivistic lines, not unlike Islamic science. It would have lacked the incentive to develop as an ideology in its own right, rivaling the transcendental worldview of the Medieval Church, proclaiming the power of nature when aided by the human mind.

Modern science reflects all the historic forces that were casting their shadow over its birth: the technological impulse of the Germanic nomads and early settlers, the founders of Medieval society; the concern for the empirical detail, the innate positivism and this-worldliness — as well as the internationalism — of the Medieval businessman; the early capitalist's bent toward intensified productivity; the drive for relentless practical progress born from the obstacles the feudal system was constantly throwing in his way; and, lastly, the philosophical dimension imposed by the universal orientation of the Medieval mind. From market to workshop to counting house, from cathedral school to castle to city wall, the life of the Medieval town impressed itself on the emerging body of modern science.

On the other hand, if the Medieval scene was marked by strident contrasts, it fell mostly to science — in conjunction with philosophy — to devise a scheme for intellectual reconciliation. Ultimately, modern science arose from the need of Medieval people to reconcile the realities of a burgeoning economic prosperity with the world-denying traditions of the Medieval mind.

* * *

Yet, though the historical constellation was uniquely fortunate, it was, in fact, anything but fortuitous. Obviously, no one could have planned this remarkable coincidence of diverse historic forces — nor the countless stimuli they radiated on the creative mind. Science as a creative talent found itself responding to unending incentives. The hour was there. The young giant was stirring. Depending on talent and temperament, a person might feel intoxicated with the fantastic opportunities emanating from workshop or cathedral school — or else sense amorphous dangers.

During the thirteenth century, both of these anticipations found their eminent spokesmen. Roger Bacon, the great Franciscan teacher and thinker, from a profound insight into the potentialities of the scientific method foresaw an age of science, some six or seven hundred years to come, with a sharpness of vision that is stunning in its prophetic detail.* Around the same time, Thomas Aquinas raised his warning voice, predicting that the unchecked growth of the kind of rationalism the new science engendered might end by alienating humanity from God's universe, and therewith from itself and from life. Though both men have in essence been proven right, they found unequal responses: Bacon was jailed by his order; Aquinas (after some controversy inside his own order, the Dominicans, because he advocated the new rationalism, albeit with certain restraints) was beatified and finally sainted.

---

* In his *Epistola de secretis operibus,* chapter 4, Bacon wrote: "Machines for navigation can be made without rowers so that the largest ships on rivers or seas will be moved by a single man in charge with greater velocity than if they were full of men. Also cars can be made so that without animals they will move with unbelievable rapidity... Also flying machines can be constructed so that a man sits in the midst of the machine turning some engine by which artificial wings are made to beat the air like a flying bird. Also a machine small in size for raising or lowering enormous weights... Also machines can be made for walking in the sea and rivers, even to the bottom without danger... These machines were made in antiquity and they have certainly been made in our times, except possibly the flying machine which I have not seen nor do I know anyone who has, but I know an expert who has thought out a way to make one. And such things can be made almost without limit... and mechanisms, and unheard of engines."

As an intellectual system, Western science began neither in the workshops nor in the marketplace. It developed in the seats of Medieval learning, which by the twelfth century were the cathedral schools. It seems no more than logical that science in this environment should begin in the form of a natural philosophy — in the abstract terms of cosmological thought consistent with the philosophical traditions of the Medieval mind. But by a strange twist, not untypical of the fabric of Medieval culture, these weighty, intellectual developments took place in a lovely and fascinating setting.

THREE

Science and Faith at Chartres

One generation passeth away,
and another generation cometh:
but the earth abideth for ever.
Ecclesiastes 1:4

T HE TRAIN FOR CHARTRES leaves from the old Gare Mont-
parnasse at 11:10 in the morning.* It glides through the
outskirts of Paris; later past the characteristic pastel shades of
the northern French countryside — woods, farms, and villages —
toward Versailles and on. It then enters a great rolling plain, the
bleached white of wheat fields stretching as far as the eye can see.
At last the tip of a giant finger emerges above the horizon, age-
blackened against the bleached white of the wheat, growing taller
before the relentless onrush of the train, until it reveals itself as
the spire of the cathedral of Chartres.

It was a clever idea on the part of the French railroad management to have scheduled the train's arrival precisely at the stroke
of twelve. This way, the cathedral bells begin tolling noon at the
exact moment the train pulls into the little station, enveloping the
passengers with their solemn chords as they spill out onto the platform, and then onto the Place de la Gare. Within a moment, the
station building, the provincial railroad, all the contraptions of a

* In recent years that ramshackle old railroad station has been replaced by a
supermodern structure, housed in an office high-rise and equipped with the
latest automated gadgets for ticket-selling and other such functions. But these
sober beginnings have done nothing to diminish the spell of the journey itself.

civilization absorbed in its trivial efficiencies, have dwindled to the size of toys. The bells are sweeping the soul clean of all its accumulated modern irrelevancies, baring it to the voice of an age bent on more fundamental matters. Towering above the sleepy little town, the great Chartres cathedral finds the traveler oddly defenseless before the message tolled by its bells.

Nevertheless, the cathedral of Chartres does not stand merely as a monument to an age attuned to the stirrings of the spirit. For its contemporaries, the cathedral heralded the future, much as to us it now symbolizes the Medieval past. Rising above the chestnut trees and the dark gray or whitewashed houses of the little French provincial town, the cathedral, historically, symbolizes the earliest beginnings of our own age of technology and science — even though such beginnings were still very much steeped in the transcendental profundities of the Medieval mind.

The great structure itself, for one thing, represented an enormous step forward in our mastery of the laws of statics — the construction of exceptionally tall buildings with a precarious distribution of weight. One of the earliest Gothic cathedrals, its bold pointed arches, its very height, its artful distribution of the masonry held up by the tension of arches and ribs and vaults, appeared as nothing short of a miracle of technology to the generations that saw the building go up. Even while the first portions of the structure were rising from the ground, some staggering new scientific insights were being formulated at the School of Chartres, in studies and classrooms housed in neighboring buildings, and eventually inside the cathedral itself. In time, an intellectual structure was to arise from these foundations that was hardly less impressive a manifestation of the power of the Western mind than the Gothic cathedral itself.

For it was at the School of Chartres that the philosophical groundwork was laid for the rise of Medieval and early modern science. Here the study of nature was established as a discipline in its own right, unhampered by older doctrinaire restrictions; here a conceptual seed was sown, from which the plant of Western science was to sprout forth into its full-blown growth and into all the branches of modern specialization.

At Chartres during the twelfth century the study of science was

The cathedral of Chartres is one of the earliest Gothic cathedrals that were built in an approximate 200-mile radius around Paris, mostly during the twelfth century. Its lofty, imposing effect was combined with a major technological achievement.

first given a definite priority over the teaching of the liberal arts, and professors advocated bold reforms for higher education as a whole, centering the curriculum on the natural sciences of the quadrivium — arithmetic, music taught largely as a mathematical discipline, geometry, and astronomy — rather than on the traditional humanities of the trivium — subjects then called grammar, rhetoric, and logic. Here the exponents of the new, scientific view of the world had to face the outraged denunciations of their more conservative colleagues at the great cathedral schools of Orléans, St. Victor of Paris, and Laon. At Chartres the writings of ancient scientists were systematically collected into a first library of science for the Western world, a basic book of knowledge from which the masters of Chartres could draw their inspiration and develop their original thoughts, which future generations were able to expand.

Behind the geographic breakthrough of the Renaissance lay the pioneering work of three centuries. Science had to be established as a legitimate area of systematic study in the Western world; the fundamentals of the scientific method had to be developed; a beginning had to be made in placing all new scientific thought

The School of Chartres was probably housed in portions of the cathedral and surrounding buildings, like this wing — of more recent design — flanking the cathedral's peaceful backyard.

upon the solid foundations of ancient science; the laws of the cosmos had to be perceived, in however tentative a way, before the shape of the earth could be explored in any more explicit detail. The enormous spadework of Medieval science, the West's first vigorous steps toward the conquest of nature, was initiated at the School of Chartres in the twelfth century.

The wrath of the religious conservatives, which the pioneers of Chartres attracted upon their heads, had its understandable reasons. For a good seven hundred years, nature had been presented as the passive object of God's creation, devoid of any innate powers to create by itself. Now the masters of Chartres were asserting that nature possessses intrinsic creative powers that were unfolding according to inherent laws or patterns of their own and whose investigation, they insisted, was a perfectly worthy subject for the human mind. Seven centuries of Christian teaching about the place of nature in the scheme of God were being challenged at the School of Chartres.

It was as though a veil under which nature had lain asleep for seven hundred years had suddenly been torn away. Even more alarming, the human race had been brought into a new direct contact with nature, implying untold dangers not only for the sleeping princess but for the gallant knight himself. Under the magic touch of Chartres's new theories, nature was stirring and waking up to life. Traditional minds were understandably frightened by the prospect of nature unleashed and humanity involving itself in her capricious powers.

Evidently, these early discords contained the seeds of the historic war between theology and science, which was to plague the growth of Western science through the later Middle Ages to the Scientific Revolution, through the trial of Galileo into our time.* And yet the roots of that historic conflict were not as deep as it may seem. What may look in retrospect like an irreconcilable dichotomy between "science" and "faith" (or "reason" and "reli-

* However unconsciously, these early (but remarkably strong) reactions also implied elementary fears of nature's unleashed forces that would seem perfectly understandable to us. Historically, however, Medieval opposition centered on the human attitude toward nature, alarming the more mystical traditionalists by the bold rationalism displayed by the masters of Chartres. It was this that made the problem seem to lie in the religious sphere. (There may have been a deeper philosophical justification to these fears.)

gion") began as a mere conflict between two ways of understanding the religious universe. These early conservative critics of the new scientific world-view were really little more than just that — conservatives, people unable to adjust their thinking to new insights and ideas, typical representatives of the inevitable slowness of the human mind.

Nor did it occur to the masters of Chartres to sever the natural universe from God's world. In their vision the laws of nature, the perceptions of the mind — as much as the contributions of the ancient philosophers to scientific understanding — were all encompassed within the divine universe and its design. Chartres cathedral and its statuary stand as a visual manifestation of an abiding conception of the universe, spanning the past and the future, nature as well as faith, Christian religion and scientific thought, the world of the Bible and the ancient world of Greece and Rome, the teaching of the liberal arts and the teaching of science — a tangible embodiment of the spirit that pervaded the Chartres school. Ptolemy the cosmologer, Pythagoras the mathematician, Aristotle the teacher of exact rational thought and of the systematic order of the scientific disciplines — all sit beside Christ and the saints, together with the founders of the liberal disciplines, on the beautiful tympanum of Chartres's Royal Portal.

It was not the spirit of newly born science that rebelled against faith. It was the timorous pedantry of conservative theologians, committed to a more limited view of God and the world, that eventually forced science on the defensive: the universe of these traditionalists was simply not large enough to contain both science and faith, nature and the Good Lord. Conservative theologians from Paris, Orléans, and Laon, prodded by St. Bernard, that ubiquitous Medieval conservative, were hounding the masters of Chartres, summoning them to appear in tribunal, denouncing their science as heresy, branding the teachers of science as rebels. And from that moment, the conflict was on.*

* * *

* By the thirteenth century, Thomas Aquinas was to show through trenchant philosophical argument that scientific rationalism and empiricism are perfectly compatible with a mystic or religious conception of the world, as long as rationalism remains aware of its metaphysical limitations. That the historic conflict did not end then and there may prove that it was largely nourished by mutual misunderstandings.

Nowadays, visitors wind their way up the steep cobblestone street, past Medieval houses, to a broad plaza on top of the hill, from which the eye sweeps clear across the wide country. But the mighty cathedral that crowns both the hill and the town is only in part the same structure in which the twelfth-century masters were arguing about problems of natural philosophy. Toward the end of that century, in 1194, most of the cathedral was consumed by a terrible fire. One can imagine the awe with which the good townspeople stared at the holocaust up there on top of the hill (and perhaps the glee in the eyes of certain clergymen who may have felt about the school much as did their brethren at Orléans and Laon) as God's wrath appeared to punish this seat of frivolous learning; as those intruders on the innermost secrets of the world now seemed to receive their just due from the very forces with which they had tampered.

But once the raging flames had died down, the cathedral was swiftly rebuilt. The famous structure we see today is mainly the product of thirteenth-century builders working from the blueprints of the twelfth. The beautiful stained-glass windows — miracles of a craft the Middle Ages refined to such perfection that the devout literally believed that to glance through them was to glance into heaven — were largely added during the rebuilding of the church, most of them donated by the local guilds. So were the flying buttresses, the huge outside supports that hold the façade between them somewhat like the spokes of a wheel. The delicate task of rebuilding the cathedral around some surviving portions of the twelfth-century structure became a factor in developing an important feature of Gothic technology.

The great cathedral is essentially the realization of an original concept, thanks to the conscientious work of a later age. But the solemn interior, losing itself in the somber heights of the nave, its dimness lit up by the stained-glass windows, with their play of brilliant colors contrasting with the dullness of the stone, surely epitomizes the vision of the original builders .

* * *

Among the portions that were spared by the fire was the west façade, containing the Royal Portal, with its sculptured representations of the seven liberal arts (including the natural sciences of

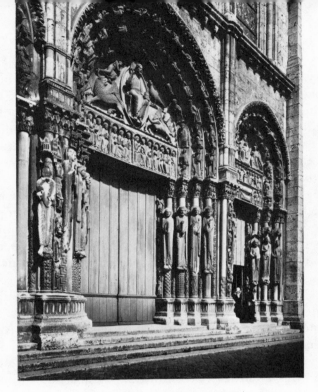

The Royal Portal of Chartres Cathedral (*above*), on its right-hand tympanum (*opposite*), contains the figures representing the seven liberal arts, like Euclid sitting under a representation of Geometry (*below*).

the quadrivium). Each discipline is represented by a pagan teacher — Donatus (or perhaps it is Priscian, another Latin grammarian), Aristotle, Cicero, Euclid, Ptolemy, Pythagoras, and Boethius (who, indeed, is the only Christian in the lot). This was an outspoken tribute to the Greco-Roman tradition of learning that was cultivated at the school. The construction of the west façade, back in the 1140s, had been supervised by Thierry of Chartres, one of the pioneers in the study of natural science, who had been appointed chancellor of the school in the same year, 1141, that the building of the west front got under way. Thus, the figures of the seven liberal arts on the Royal Portal may be taken as a visual expression of Thierry's program — or of Chartres's basic educational philosophy.

Chartres's concern with the modernization of academic teaching was mainly a question of adjusting traditional Medieval education to the needs of the dynamic society of twelfth-century France. That shining piece of French soil studded with famous old towns which surrounds Paris within a sixty-mile radius and in which Chartres is located, the Ile-de-France, was really the heartland of everything new and progressive inside Medieval society. Not only the Gothic cathedrals with their stunning technological feats and the study and teaching of science had their beginnings here, but also the early capitalistic system, which no doubt supplied the economic incentive for all this spectacular cultural activity.

The Ile-de-France was the natural trading center for Flemish and English wool about to be shipped to the ports of the Mediterranean; and towns like Chartres, St.-Germain, Rheims, Laon, Compiègne, and St.-Quentin were flourishing with the boom of the wool trade. Their early capitalist wealth reflected itself in the high-rising Gothic cathedrals — symbols of their civic pride. Their militant feeling of independence expressed itself in the organization of the communes, political groupings designed to wrest power over the towns from the feudal nobility. The same spirit of independence motivated the educational reforms, with their accent on natural studies, expressing some of the mental directions of the new commercial society.

Chief spokesman for educational reform at the School of Chartres was William of Conches. Together with Thierry and another master of Chartres, Bernard Sylvester, Conches was also a pioneer in hammering out the new natural philosophy, the foundations of

Western science. As a matter of fact, Conches's fight for revising the academic program was a logical outgrowth of the importance science played at the school. But the statuary of the Royal Portal, placing all seven arts on an equal footing with each other — and on a level with Christ and the saints — shows that the new emphasis on science sprang from a thoroughly humanistic concept of the intrinsic unity of the world, in which the world of ideas and the world of nature were seen as one. The human being and the world of nature were of one piece in the thinking of Chartres; the social sciences of the trivium and the natural sciences of the quadrivium were both studied as aspects of a single universe. For the masters of Chartres, an understanding of nature was part of what they called *humanitas*, the true humanity of a cultured individual.

Today, we, who are caught in the backwash of the same historical process that was released then and there, tend to feel that a touch of humanistic education might create a wholesome balance for the human and cultural one-sidedness of the overspecialized scientist or engineer. The pendulum has swung in the opposite direction. Of course, we have to cope with a body of scientific information swelled to gigantic proportions by the work of some eight hundred years, while the Chartres masters were merely helping along the emerging study of science. It is nevertheless a pleasant thought that the beginnings of science-teaching in Western civilization sprang from a concept of the intrinsic unity of all knowledge, rooted in a calm faith in the essential unity of the universe itself.

* * *

What kind of science was it that was cultivated at twelfth-century Chartres? Evidently, the vast edifice of Western science (we call it "modern science" only from the sixteenth century on, habitually disregarding the spadework of those first four hundred years) was merely rising from its foundations, and the foundations themselves had first of all to be laid. The decisive step taken at Chartres was the formulation of a "natural philosophy" — the establishing of the basic philosophical premises within which Western science was then able to develop and branch out. Before any explicit investigation of nature (and, with that, the unfolding of specialized sciences) could get under way, certain fundamental premises had to be clarified. Chartres's main contribution was that it did precisely that. Yet the thinking at Chartres revolved by no means around abstract constructions. What inspired these men was a perfectly concrete vision of nature, a fresh and vital view of the natural cosmos (or a distinct cosmology, as Medieval scientists would have called it), leaving it to the future to fill these cosmic outlines with the respective astronomical, physical, mathematical, medical, biological, chemical, geological, and geographical detail.

Nature till Chartres had been essentially alien to the Medieval mind. The alienation stemmed from the otherworldly mentality, with its denial of the world of the senses, that had been proclaimed by St. Augustine back in the early fifth century, in the very twilight of ancient Rome. Suited to the catastrophic condition of the West during the collapse of Roman civilization (and just as suited for the harsh frontier life of the early Middle Ages), it was this kind of intellectual heritage that the masters of Chartres sought to reverse. They were in part responding to the social conditions of the new early capitalist prosperity, while their more conservative colleagues at the neighboring cathedral schools upheld the traditional, strictly otherworldly interpretation of the Christian faith.*

As a further important step toward reality, Chartres took the lead in reconstructing the scientific knowledge of the ancient world, thereby establishing a firm basis for the coming evolution of Western science. The turn toward nature was accompanied by a deliberate turn toward the foundations of the past. What this involved, in concrete terms, was the systematic expansion of the still appallingly meager library of ancient scientific texts available in Latin (part of a general effort to recover Roman literature,

* We tend to think that it was the Church as an institution that was blocking serious scientific progress (if not all serious rational thought), but such modern generalizations amount to oversimplifications. Chartres itself was evidently a Church-sponsored cathedral school. The teachers were frocked and robed members of the clergy. The new initiative for the study of science was developed under the protecting hand of one of the most respected bishoprics and cathedral chapters of Medieval France. Since the Church, way into the high Middle Ages, exercised an almost total monopoly over the intellectual life, the situation could hardly have been otherwise. Virtually any important new intellectual movement had to originate in some quarter of a not-as-yet fully centralized (or dogmatically monolithic) Roman Catholic Church.

which was explicitly cultivated at Chartres); together with a vigorous initiative toward the translation of scientific texts from the Arabic and their use in teaching and study — a momentous contribution to the rise of Western science, about which much more will be said. Clearly, the spirit of humanism that we like to associate with the Renaissance was already very much alive at the cathedral school. One was aware at Chartres that the barbaric West needed the soil of classical thought to nourish its own creative thinking, in philosophy and literature as much as in science. The beautiful ideal of *humanitas* included this kind of historic awareness.

Humanistic culture at Chartres reached its fullest flowering during the 1170s, a generation after the renaissance of scientific studies. The classical purity of the Latin prose, the familiarity with the great Roman authors, the whole educational philosophy that prevailed under John of Salisbury's chancellorship belong to the peaks of the classical tradition in the Western world.

The impulses of the cathedral school were to vibrate far into coming centuries. The men of the Italian Renaissance who reconstructed the globe with the help of Ptolemy and Strabo were decidedly indebted to Chartres's initiative in reviving classical culture in general, for the first deliberate steps in the reconstruction of ancient science, and for the first forceful impetus toward original scientific inquiry itself. A consistent tradition of active scientific thought, a continuous effort in the regeneration of classical science, inspired by a deep reverence for the classical past, leads from the School of Chartres to the Renaissance and the age of discoveries. It was under the Gothic vaults of the cathedral of Chartres that the long process of the return to nature was begun.

Formulating the philosophical premises; defining the basic concept of the cosmos from which all later specialized sciences were to grow; systematically reconstructing the scientific knowledge of the past and thus placing the coming evolution of Western science on a solid traditional footing — each one of these steps seems so crucial that, taken together, they could only mean one thing: that in a period of fifteen to twenty years, around the middle of the twelfth century, a handful of men were consciously striving to launch the evolution of Western science, and undertook every major step that was needed to achieve that end. We are facing one of those rare moments when a movement of enormous historic consequence is initiated with perfect awareness — and very nearly total success.

Still, it is strange to think that all this purposeful activity on behalf of the age of science should have taken place at a site that to us seems the very embodiment of Medieval faith. The abiding fact is that modern science grew out of the lovely Medieval idea of *ordo mundi*, the faith in a universal order, a religious feeling for the ultimate unity of all life.

*   *   *

The recognition of nature as an autonomous, largely self-motivated world — the crux of Chartres's natural philosophy — was a startling new concept for the Middle Ages. The reasons are not so much theological as philosophical. It was not simply a question of the priest discouraging the sensuous enjoyment of nature with threats of punishment in purgatory or damnation in hell. For seven hundred years Medieval thought had dwelled on the infinite regions of the beyond, looking down on earth as a sphere of minor significance, in fact of lesser reality than the great invisible "universals." That other world, for traditional Medieval thought, was not simply the afterlife, something that would take on reality only after one's death. It was there simultaneously with this life, as a higher, albeit invisible, sphere to which one could lift one's spirit at any moment, from which one received comfort in sorrow, clarification amongst one's confusions, meaning during one's chaotic existence down here. From the peasant to the philosopher, Medieval minds had been trained to look on this life as if with the bird's-eye view of that higher region, where all ideals and values — the universals — had their eternal home. Medieval philosophy had invested a great deal of sharp-edged thought in exploring the workings of that timeless dimension, tracing the invisible threads that were linking it to the limited world of human affairs. There was much profound wisdom, much to sustain one's inner security in that cultural vision, which modern Western civilization appears to have lost. Yet the natural world seemed merely a distant region of shadows. To the mind trained on eternal visions, nature seemed only a darkling patch.

A mental habit developed for many centuries, encouraged by

acutely unfortunate living conditions, sanctioned through theological argument, will hardly be broken overnight. It lingers on to form the framework for new ideas, the obsolete frame of reference for any fresh perception or experience that life might bring.

St. Augustine, chief architect of Medieval thought, had assigned nature a pathetically insignificant place in the great otherworldly context by postulating that the Good Lord, in creating the world, had left behind what Augustine called "seeds of causation" (rationes seminales). He visualized these as actual physical entities — bodies "of a humid nature," as he said. Conveniently, that picturesque concept came to account for all the observable facts of birth, growth, and evolution (what the Middle Ages liked to call the phenomena of generation), which were supposed to be functions or qualities of these causal seeds. The whole realm of natural evolution, in other words, was left a mere by-product of the act of God's creation. Whatever else was to happen in the natural world after the six days of creation was implicitly of minor significance, hardly worthy of detailed observation or explicit thought, certainly not the result of complex inherent causative patterns or independent evolutionary laws. In a way, St. Augustine's causal seeds had turned all of nature into a charming — if somewhat neglected — little cloister garden in which, with the Good Lord's benevolent help, things were left to take care of themselves, while man could devote his mind to loftier, more spiritual matters.

The Chartres naturalists were not content with this pious explanation. As they saw it, nature was a great deal more bountiful and far more replete with miraculous innate powers than Augustine's seeds-dropped-in-the-garden concept seemed to imply. Creation had not ended with the six days; the natural world was clearly the scene of a creative process continuously, endlessly, in infinite variations, still going on under one's eyes, if one chose only to look.

Basically, Chartres's theories (and those of twelfth-century naturalists in general) all flow from this new focusing of the Medieval vision on nature as a vital, continuously creative force — an "insight," in the literal sense of the term, like people opening their eyes to some familiar situation whose real essence they have failed to notice before. The School of Chartres even sparked a literary movement, from the French poets Alain de Lille and Jean de

Meung to the Italians Dante and Petrarch, which turned this new vision into often glorious poetry. At Chartres the very perception had the rank of discovery (modern writers have called it a "discovery of nature"), and the school's main achievement was in exploring the precise implications of the new insight.

One of the first tasks was the reinterpretation of the Book of Genesis. If the new vision contradicted the doctrine of a finite series of creative acts completed within the six days, the Good Book had to be read largely as a symbolic story involving the positive concept of natural creation as an evolution that had not been concluded but was still very much going on. In his lectures on the first chapters of Genesis, Thierry of Chartres re-explained creation "solely in terms of natural causes" (*iuxta physicas rationes tantum*), as he explicitly said. The same man who, as chancellor of the school, initiated the statuary of the Royal Portal symbolizing the unity of science and faith, taught in his lectures that the story of creation in the Bible was compatible with a scientific approach. It was a startlingly modern interpretation that would have shocked the fundamentalists of modern times. Thierry was, in fact, denounced by some outraged contemporaries as a "magician" — whether by his colleagues or students (or their parents), we do not know.

Thierry's theological basis, besides the reinterpretation of Genesis in these naturalist terms, was the idea of the gradual "beautification" or "decoration" of the world by God (which some of the Fathers of the Church had already begun to distinguish from the mere act of creation itself) and which, obviously, implied a continuous process of evolution of some significance. His lectures, collected in book form, became the groundstone of Chartres's natural philosophy, and Thierry was worshiped by a generation of students who imagined that Plato's soul had been reincarnated in him.

It was also Thierry (together with William of Conches) who introduced certain fundamental Platonic and Pythagorean ideas of the cosmos to the Medieval West — above all, a basic vision of the ordered structure of the universe, which was to live on as a vital idea in Western thought, as an inspiring concept for the Renaissance as well as the Scientific Revolution. It was Thierry of Chartres who gave Western science its basic conceptual framework through a liberal interpretation of Christian teachings, enlivened by the Platonic vision of the cosmos. The foundation of science in a philosophical concept of an ordered universe goes back to the same man who was evidently the main inspiration behind the development of the whole school.

It was Thierry as well — under whose chancellorship in the 1140s the school attracted students from all over the Western world and took on a distinctly cosmopolitan hue — who first stimulated the search for ancient and Arabic scientific manuscripts in Spain. The result was that some striking Aristotelian and Islamic ideas appear sporadically, but unmistakably, in the thought of Chartres.

We are left to draw the portrait of this remarkable man from inferences based on his theoretical writings and the enthusiastic comments of his students and can therefore only try to guess at the quality of his mind. The earliest pioneers of modern science do not stand before us with the flesh-and-blood realism of a Copernicus, a Galileo, Descartes, or Newton. Their portraits are dimmed by the great distance in time. But the mind of this "Plato reincarnate," as we perceive it, seems radiantly lucid in his beatific vision of the rational order of the universe. We may smile, too, at his sarcastic glint as he supported his predecessor in the chancellorship of the school, Gilbert de la Porrée (whom he did not mind attacking hotly in theoretical matters), in denouncing the materialistic mentality of some Chartres students whom Master Gilbert advised to take up the baker's trade instead of bothering with the thankless intricacies of serious academic studies.

On the whole, this twelfth-century chancellor of the cathedral school sounds amazingly modern: modern, the gently ironical teacher amidst the international student crowd, youths eager to grab their slice of the prevailing commercial prosperity; modern, his rational interpretation of religious doctrine; curiously modern, his idea of a continuous creation — which evidently underlies his concept of God's continuing beautification of the world; modern, his dynamic leadership, with the impressive diversity of his projects and the vigorous impact of his initiatives; modern, his vision of the ordered cosmos, anticipating the esthetic philosophy of the Renaissance and René Descartes's *mathesis universalis*. But, then, the retracing of scientific thought has a way of revealing unexpected intellectual kinships with distant cultural environments, making the Medieval world seem all at once more "modern" (and perhaps ourselves more "Medieval") in a sudden surprising flash.

Thierry's mind excelled on every level where Chartres was to make its historic contribution — disentangling nature from a narrowly fundamentalist interpretation of faith; sketching the philosophical outlines of a scientific cosmology, as a starting point for more specialized investigations; building a solid foundation in the stored-up scientific knowledge of the past. Thierry stands out as the great initiator for the advances of Chartres, forward-looking but still able to contain them within the traditional universalist framework, under the shelter of his powerful chancellorship. Some day Thierry will probably be recognized as one of the true founders of Western science. The great vision of the harmony of science and faith on the Royal Portal is certainly a worthy tribute to the quality of his mind.

Our age, suffering under the increasing separation of science from the overall context and meaning of life, may have cause to admire this early pioneer, who courageously advanced scientific thought in the face of his outraged critics yet still preserved his vision of nature as merely one aspect of the world.

* * *

The next, and probably the greatest, of the Chartres naturalists strikes us as a man of very different fiber, although he grew up in the same intellectual climate and belonged roughly to the same generation as Thierry. William of Conches's voice still rings defiantly across the ages as he hurls back at the bigots who were accusing him of blasphemy: "I take nothing away from God: He is the author of all things, evil excepted. But nature with which He endowed His creatures accomplishes a whole scheme of operations, and these too turn to His glory since it is He who created this very nature."

What a proud voice! And what a vision in which nature now stands fully in her own right, accomplishing "a whole scheme of operations," and yet redounding to God's greater glory! No longer St. Augustine's causal seeds dropped in the garden. No more entanglement with, or disentanglement from, the whole burden of the theological legacy. Instead, the positive vision of nature as a creative process, coupled with the clarion call to the human mind for unhampered scientific research (which, as Conches made clear, must follow from the recognition of nature as an autonomous dimension). No wonder it was Conches whom the conservatives singled out for their fiercest attacks and denounced as a heretic.

Forced to resign from his teaching position at Chartres, he returned to his native Normandy (he was born near Evreux), yet remained unbroken in spirit. Dialectical, outspoken, thriving on controversy — in a word, ruggedly French — he at last achieved his compatriots' due recognition. A few years after the publication of his chief work — a revised version of his *De philosophia mundi,* which he called the *Dragmaticon* — a Norman chronicler mentions him proudly as a famous man living in the neighborhood.

* * *

Conches made his impact on the history of education almost as much as on scientific thought. John of Salisbury, who had been his student and became a leading educational reformer in his own right and time — the twelfth century's outstanding spokesman of classical culture — acknowledged Conches's influence with profound respect. The admiration he repeatedly expresses in his writings for his teacher is all the more remarkable, considering that John of Salisbury, with his own humanistic educational philosophy, was quite critical of the heavy emphasis on scientific studies that Conches had advocated.

In contrast to Thierry's prodigious energies as administrator and initiator of new projects and general ideas, Conches shows the vigor of radical reform and original thought. If Thierry knew how to weave the new insights harmoniously into the traditional texture, Conches provoked storms of controversy through the outspoken positiveness of his mind. His was the kind of mind that alarms and causes trouble; also the kind that makes history by its uncompromising force.

Conches's system of nature was distinguished by three principal features: first, by the very fact that it represented indeed a consistent system; second, by the dynamic conception on which it was based; third, by his singular grasp of nature's essential autonomy, which, as he realized, made it accessible to the rational mind. Each one of these aspects was of fundamental significance.

That this twelfth-century thinker was not satisfied with dashing down some more or less unconnected new ideas, but had to work them into a coherent new system, shows the unmistakable imprint of the spirit of Chartres as it had been visualized by Thierry: the universalism of Chartres demanded to see the world as a unified whole. Where Thierry's universal vision remained in essence a matter of general conviction, Conches tried to replace the otherworldly universe of the conservative theologians with a new natural cosmos that was to be equally complete and at the same time this-worldly, though still the unquestioned handiwork of God. In this sense, it was chiefly William of Conches who established the theoretical framework of Western science, which in every phase of its growth has shown a distinct tendency to relate all specific observations to some sort of consistent overall context. Whenever scientific thought had sporadically appeared before, since the days of Gerbert of Rheims around the turn of the millennium, it had been notoriously lacking such a coherent frame of reference, nor did it have the universal vision that would place it in the realm of natural philosophy. Only when Conches decided to spell out Thierry's conception did specialized scientific thought attain the rank of philosophical consistency.

Thierry's concern had been chiefly with the balance between Medieval tradition and fresh scientific research. His major contribution had been in the nature of an effective rearguard action against a fundamentalist interpretation of faith, thereby freeing the way for science. William of Conches created a positive system of nature. Conches's conception was systematic in that his universe made sense in its own right, according to nature's inherent laws. As far as his writings go into theological questions (which they do mostly by reviving Greek thought within a Christian religious framework), his approach is always emphatically naturalist and ultimately philosophical. His purpose is no longer to justify the study of nature through subtle theological argument (as had been Thierry's), but to show that nature functions consistently, with God's acting both as "first cause" and as the prime force behind evolution — the ultimate principle of causality and the perennial source of all life. With a leap, Conches's vision anticipated the philosophical debates of the Scientific Revolution, during the seventeenth and eighteenth centuries, which once again confronted the problem of reconciling a scientific conception of the natural universe with a basic deistic belief in God.

★　★　★

Yet where the philosophers of the Enlightenment were able to think of the highly explicit cosmos of early modern science, William of Conches had little more than the first hazy outlines of the natural universe in mind. (Even the Aristotelian cosmos was still essentially unknown in Conches's time.) For all his emphasis on scientific study, Conches's vision of an autonomous nature, created and continuously developed by God, was nothing much more than a vision — the product of almost pure insight. Like Moses on his mountain glimpsing the Promised Land, Conches recognized the reality and essential autonomy of the natural world, though its detailed features were as yet veiled behind the mists of time.

Conches's concept of nature was at the same time "dynamic," because he perceived of nature in a vivid and tangible sense as motivated by its own inborn creativity. In his mind, St. Augustine's causal seeds took on a distinct life of their own. The distant vision of the natural universe was vibrantly alive.

Conches is so completely explicit on this point that its importance for the evolution of Western science deserves to be clearly understood. To the modern view the Medieval cosmos appears like a "closed" — and, by the same token, invariably static — world. Only the Renaissance seems to have replaced this closed cosmos with the modern idea of the infinite and dynamically changing universe (somewhat tentatively through Nicholas of Cusa in the fifteenth century; more definitely through Giordano Bruno in the sixteenth). Historians are therefore surprised to find that a dynamic conception of a continuously evolving — and, hence, continuously changing and expanding — natural universe existed already by the thirteenth century among men like Witelo or Roger Bacon, and that it pervades the cosmic vision of some of the great fourteenth-century physicists. Yet as soon as one tries to visualize the Medieval idea of nature in concrete terms (rather than quibbling over fine semantic points), it turns out that Conches at his twelfth-century school already saw the universe as forever evolving and growing — as dynamic a vision, in principle, as that of any modern astronomer, physicist, or biologist.

Our modern impression that Medieval science saw the cosmos as static is probably based on Aristotle's commanding influence, which brought an almost changeless cosmos, turning eternally on its concentric spheres, to the later Middle Ages. Yet Aristotle, even after his cosmology had been revived, never did achieve that total control over Medieval science that we tend to attribute to him. There seems always to have been a native Medieval current, presumably mystical in its origins, that preferred to see nature as full of its own capricious life. When Medieval culture learned about the Aristotelian system through translations from the Arabic, half a century after Conches, this native current soon generated a proliferation of severe criticisms of Aristotle, which finally ended by destroying the Aristotelian cosmos. Meanwhile, from Conches through Nicholas of Cusa to Bruno, the native dynamic idea of nature was kept alive.

The dynamic motif was no doubt the most modern element in Conches's concept of nature. A master of the Chartres School with a more poetic mind, Bernard Sylvester, went so far as to picture nature as a pagan goddess, the embodiment of eternal fertility, *mater generationis,* the forever-procreating mother-goddess, Venus herself. Such poetic expressions (foreshadowing the feeling that so decisively pervades Renaissance art) were to mark the first stirrings of the modern world's love affair with nature — perhaps the strongest emotional motivation behind modern science — breaking forth from the wellsprings of the Medieval mind.

Conches spoke in more sober accents, although no doubt from the same sense of discovery — a feeling for the poignant loveliness of nature's eternal vitality. He detected "causes" and "second causes" and perceived that even what he called the second causes are forever carrying forth the creative process. Conches saw, too, that nature creates according to miraculously prefigured patterns, so that "like things are brought forth by like things, out of the seed or the bud, as nature is a force that is inherent in [all] things generating like things from like patterns." Although he still had the Medieval capacity for wondering about the elementary phenomena of life, he recognized that they occurred in patterns of unending repetition, within every single detail of the natural universe. Behind his causes and second causes was, of course, a deep feeling for the miraculous phenomenon of causality. And behind his wonderment at nature's creative patterns, inherent in "the seed or the bud" (or the chromosome or the human fetus, we may add), was a marvelous sense for the way nature acts through archetypal patterns containing the complete potential of any subsequent growth.

Precisely because Conches possessed the naive ability to wonder about elementary processes we tend to take for granted, this twelfth-century Frenchman probably had a keener sense of the phenomenon of evolution than do most of our modern scientists.

* * *

What is more, on a stage no longer second-rate in importance, the human mind itself had become the protagonist. "To seek the reason of things and the laws of their origins," Conches wrote, "is the great mission of the believer, which we must carry out by the fraternal association of our inquiring minds. Thus, it is not the Bible's role to teach you the nature of things; that is the domain of philosophy." If Thierry had attempted to read the Book of Genesis in the light of natural causes, Conches went beyond the religious context, establishing the investigation of nature as the legitimate realm of "philosophy," by which he now meant the free inquiry into the natural world — or, in other words, science. He envisaged, and thereby helped launch, science as an independent collective enterprise, an effort carried out "by the fraternal association of our inquiring minds," an intellectual community of the kind he unquestionably had experienced at Chartres.

One perceives still a deeper meaning underneath these programmatic statements, a feeling for the strange correspondence between the workings of nature and the workings of the rational mind. The curious phenomenon of nature's appearing to proceed according to built-in "rational" laws that are susceptible to our understanding, precisely because the human mind seems to operate according to the very same rational patterns, has been considered a tantalizing problem almost since the beginning of philosophical inquiry. Yet modern science (if, indeed, not modern philosophy) has in effect chosen to ignore this philosophical problem, taking it for granted that those intrinsic rational laws of nature and the laws of our own logical thought are one and the same.* At Chartres, the apparent coincidence between the two processes was both noticed and established as a "given" for the future evolution of Western scientific thought.

We stand at the birth of the concept of "natural law," which was to pervade Western science from then on; or at the point when the ancient concept was reborn, in direct application to science. Chartres discovered (or, more exactly, rediscovered) the concept — pervading Western science ever since — that nature is intelligible for the human mind precisely because both proceed according to the same inherent rational laws.

"It is through reason that we are men," said another twelfth-century naturalist who had studied at Chartres, Adelard of Bath, with an inspired phrasing. "For if we turned our backs on the amazing rational beauty of the universe we live in we should indeed deserve to be driven therefrom, like a guest unappreciative of the house into which he has been received." Nature is beautiful as a whole because, like music, it evolves within the harmonious patterns of its innate rational laws. It is our task to grasp that beauty, grateful for the hospitality with which the Good Lord has received us in His house, by applying the powers of the rational mind.

The earthly paradise had reopened its gates, the human race had returned to the Garden of Eden, and the only sin to avoid henceforth was to turn one's back on nature's "amazing rational beauty." The esthetic delight of the mathematician — and also of the scientist working in most other disciplines — has flowed ever since from the perception of the intrinsic harmony that appears to prevail between the inner lawfulness of the natural universe and the laws that rule the rational deployment of the mind. At the cathedral of Chartres, this must have seemed like the discovery of a wondrous concordance.

In a sense, of course, it was not: Greek philosophy had already been inspired by this overwhelming perception. Yet in one sig-

* The philosophical problem is most clearly exemplified when one stops to wonder why, for example, the course of the stars should follow the same rules by which we learn in school how to draw such geometrical figures as a circle or an ellipse, which seem like mere intellectual constructions. Behind this problem, as Plato recognized, stands the much more fundamental problem of what really is "reality" — what we observe in nature, or the workings of the human mind.

nificant sense it was: historically, the Western idea of natural law carries religious overtones, echoing Chartres's Medieval key.

Greek thought, since Pythagoras and Plato, had recognized such a concordance between nature and mind; the cathedral school, its humanist vision turned toward the classical past, was aware of this strand in the Greek tradition. Glimpses of Greek natural philosophy could be found on the shelves of the school, in some excerpts from Plato's *Timaeus* (more precisely, its first twenty-one books, in a Latin paraphrase version by Chalcidius); or in the extant writings of Boethius (used at Chartres mostly as a breviary of Aristotelian, and sometimes Platonic, thought). The masters of Chartres liked to base their lectures and original papers on this slim residue of ancient philosophy of which they had knowledge, and Conches followed the custom by writing his earlier works in the form of commentaries on the *Timaeus* fragments and on Boethius.

Nevertheless, a difference exists. On paper it looks like no more than a subtle shading in the philosophical concept. In the historical reality of coming centuries it emerged as one of the most crucial differences between the essentially contemplative quality of Greek natural philosophy and modern Western science, with its relentless bent to achieve control over the forces of nature. The difference is that the Greeks (and, after them, Hellenistic thinkers) had viewed nature by and large in objective terms, as a reality existing outside the human realm. They related themselves to it, poetically or emotionally, through their pantheistic tradition, in a feeling of reverence before this sacred playground of the gods. Philosophically, the Greeks acknowledged in nature a perfect harmony and order and could think of nature as a supreme model for the human mind. They even recognized the creative principle operating in nature, the *demiurgus,* and at times thought of it in personalized terms, a kind of all-around handyman of the gods.

But the Greeks lacked that total personification of the creative principle for which every natural evolution is ultimately, whether directly or indirectly, the work of the Creator Himself. This singularly Judeo-Christian idea ruled the monotheistic culture of the Middle Ages. Instead of the active, autonomous nature perceived by the Greeks, embodied in a plethora of nature gods, the Middle Ages tended to see nature as a passive object, an "extension" of the one God (which was precisely the theological problem with which the Chartres philosophers had to wrestle). This deep-rooted idea involved a formidable implication; namely, that human beings, by the God-given powers of their rational minds, ought to be able to enter the divine secret, copying or even replacing God in His creative work. From a status of joyous independence shared with the gods, nature had been debased to a passive condition. It was a question of time — and favorable circumstances — as to when the human race would take advantage of it.

A first seed of this concept is contained in the natural-law philosophy of William of Conches, who insisted that the creative principle of the Greeks, the "demiurge," was in truth identical with God. By the thirteenth century, the idea was further spelled out in the mystical philosophy of Roger Bacon, who saw science as a kind of surreptitious effort to enter into the divine secret and so, in effect, wrest control over nature from the Creator. The idea has remained an integral (if by now largely unconscious) element of Western science: humanity succeeding God in the creative manipulation of the forces of nature.

It was an essentially religious idea in its origins, yet to the Greeks it would have seemed *hubris,* the sin of arrogance, the most outrageous disregard for the established limits of human power of which they knew.

* * *

The countryside has not changed much around Chartres since the days of the School. From the top of the hill the masters looked at the same landscape, the same rolling plain encompassed by the wide horizon, the wheat fields studded with dark green patches of wood, crowned by the chestnut trees winding up the hill toward the cathedral.

Even the town has not changed a great deal. The same little river furrowed the foot of the rock, sheltered between the backs of cozy little houses. The laundry was washed in the same waters, which mirrored it patiently when it was hung up to dry. The nature about which the masters of Chartres philosophized was the lovely nature of the Ile-de-France, changeless except for the seasons. In John of Salisbury's works, written during the late 1150s, when the fervor of the great naturalists was already spent, we can perceive a mood of weariness with the unending upheavals of a turbulent age and with the vanity of man. John was, in truth, writing in the spirit of *Ecclesiastes*: "One generation passeth away, and another generation cometh: but the earth abideth for ever." Amidst the bustle of a blatantly profit-minded and power-hungry society, the masters of Chartres had discovered in nature a new manifestation of the eternal Divine, outlasting the efforts of mortals.

There was, indeed, still a good deal of the changeless quality of the Medieval world-view about Chartres's natural philosophy, no matter how unorthodox it may have seemed at the time. Thierry and William of Conches and the other masters of Chartres had in truth spoken in a profoundly Medieval vein. In their powerful vision they had included the whole natural world within the scope of the Divine. While rejecting St. Augustine's neglect of the world of the senses, they had realized the Medieval idea of the fundamental unity of all life far more fully than their conservative colleagues. The house they built as a philosophical framework for modern science was really an inspired extension of the age-old structure of Medieval universalism, their conviction a Medieval faith in the all-pervasive spirit of God.

But how was science to proceed from there? How was it to come down from this lofty metaphysical level to the indispensable chores of systematic evidence-gathering and empirical observation? How could philosophy make contact with the real world, without having to repeat the gargantuan task of amassing data performed by some four thousand years of antiquity?

The problem was solved by a stroke of luck.

FOUR

The Gift of Islam

He it is Who sendeth down water from the sky, whence ye have drink, and whence are trees on which ye send your beasts to pasture.

Therewith He causeth crops to grow for you, and the olive and the date-palm and grapes and all kinds of fruit. Lo! herein is indeed a portent for people who reflect.

Koran, Surah XVI ("The Bee"), 10:11.

SOMETIMES history is made by luck. At the very moment when the West had worked out the contours of its natural philosophy and was ready to tackle a more detailed exploration of nature, fortune revealed a virtually complete body of specialized scientific information, without question the most complete store of knowledge of the natural world mankind had yet compiled. Luckier still, that storehouse turned out to be as good as next door to the French intellectual centers, across the Pyrenees, amidst the rugged and somber-colored landscape of Spain.

For almost four hundred years, Spain had been the scene of one of the most grueling civil wars in all history, the *reconquista* — the inch-by-inch reconquest of the Iberian Peninsula from Muslim rule. Alternating with sporadic Muslim advances, and flaring up again after periods of relative peace, the war had kept the Spanish borderlands in a state of turmoil, even as the Moorish invaders were spreading their sumptuous civilization across the heartland of Spain. By the beginning of the twelfth century, close to two thirds of the area had been won back, the Muslim forces pressed into the southern pocket below the Tagus, and many of the brilliant centers of Islamic culture occupied by Christian knights. Although the war was still far from over, though the front lines were still shifting back and forth and much of the countryside lay ravaged, peace had returned sufficiently to permit the quiet study of the great cultural heritage of Islam. A farsighted Christian ruler, Alfonso VII, king of Castile and León, had founded a center for the study of Islamic culture and science at Toledo, one of the new outposts of Christendom, where Moorish and Christian influences began to mix within the life of the streets and the architecture of the buildings, as in a marvelous tapestry. Libraries, their shelves stacked high with volumes on the most diverse subjects, awaited the scholars from the Medieval West. All they had to do was cross the Pyrenees, swarm into the former Muslim places of learning, take the volumes from their shelves, blow off the dust, and settle down to study Arabic.

That the Islamic legacy proved ready and waiting just when it was needed amounted to a lucky accident. It consisted in the coinciding of the military successes of the *reconquista* with the evolving intellectual needs of the Medieval West. Contacts with Islamic Spain had been developing gradually over the centuries. Gradual, too, had been the successes of the reconquest, but it was only after the School of Chartres had established its new natural philosophy that European scholars were plunging into the Islamic heritage with full enthusiasm. By that time most of Spain was reconquered.

The evolution of the Medieval mind and the fortunes of the battlefield happened to coincide. The rest was a question of painstaking scholarship and hard work.

* * *

Swooping down from their mountain strongholds of Navarre and León (and taking with them the name of their castle-spiked frontier region, Castile), the Spaniards had managed to regain their country from the Moorish invaders in an incredibly tough and tenacious struggle. But the country they regained was profoundly changed by the centuries of Muslim rule. Raids conducted from mountain nests, merciless fighting between two ideologically opposed camps, a country ravaged by civil war, the echoes of Muslim civilization, the colorful interpenetration of cultures — such typically Spanish elements had formed the setting in which Western science was to take its next steps.

Spain, to the intelligensia of the high Middle Ages — teachers and students or *vagantes*, wandering scholars — represented adventure. The fascination of the enemy culture that had ruled over the Iberian Peninsula had spread secretly and slowly since the tenth century at least. By the twelfth it attained the proportions of a cult.

Spain meant the glitter of the Islamic Orient, the fascination of a new brand of learning, in some ways the mysteries of forbidden knowledge. It meant a culture that could not have contrasted more completely with the still barren and ascetic Medieval world, despite the incipient prosperity of its towns. Islam had left its traces in streets and gardens and mosques, in the ceramic décor of colorful façades, in walls enlivened by horseshoe arches and delicate filigree on fountains — still splashing even though their Muslim builders were gone — in the libraries and patios of former Muslim places of learning.

To the Medieval West, Spain was like a window suddenly thrown open to the exotic life of a different world. To a culture that had been accustomed to live within its own narrow limits, the breakthrough of the *reconquista* amounted to a breakthrough into the wide-open outside world itself. The crusades (and a few less concentrated military operations) had produced scattered contacts with Islam — in the Near East, in Sicily and southern Italy, in North Africa. Military successes had opened the doors to trade. But the rolling back of the Muslim forces in Spain revealed an entire western-European country steeped in this foreign and exciting civilization.

The effect was an intellectual stimulation without parallel. Virtually every facet of European life, from religion and philosophy to governmental institutions to architecture, personal mores, and romantic poetry, was profoundly affected. For Medieval science, Spain meant the opportunity to advance in one giant step from the abstractions of philosophical thought to the tangible experience. The wealth of data offered by Islam enabled the West to fill the outlines of the new philosophical cosmos with the endless

The "Hall of the Two Sisters" in the Alhambra, Granada, displays a marvelous lacework of arabesque decorations.

detail of already well-developed specialized sciences, each embodying a rich store of experience in natural observation.

Islamic science meant more. Its abundance of concrete information, inspired by Islamic culture's love for natural detail, seemed like a fulfillment of that pantheistic vision of nature that had nourished the Chartres theorists. Islam epitomized the type of science of which the masters of Chartres had dreamed.

The steps of Christian scholars began to echo in the halls of Spanish libraries and cathedral schools, many of them former Muslim places of learning. With a fanatic fervor, a handful of scholars plunged into the study of Arabic, assisted by Spanish Jews who were versed in the language (and quite often the sciences) of Islam. At Barcelona, Tarazona, Segovia, Pamplona, León — in the Spanish North and Northeast — and above all at Toledo itself, they went to work to translate the scientific writings the Arabs had left behind on the shelves. Within little over a generation, the core of Islamic science was translated into Latin, the common language of the learned West. Within a hundred years, the West had in essence assimilated the scientific knowledge of Islam. Within less than another hundred years, during the fourteenth century, the West had decidedly overtaken Islam in its intellectual mastery of nature, thrusting ahead into the mysteries of science, with the Islamic legacy serving as groundwork.

Yet, since Islamic science was really the quintessence of the knowledge of almost all preceding cultures, East as well as West, a window had also been opened on the ancient world, a window on Medieval culture's classical past, a window on history. The flavors and colors of the Orient, the panoramic vistas of the outside world and of the past, even the smells and sights of nature itself seemed to come flooding into the ascetic studies of the Medieval cathedral schools.

* * *

Islam is one of the most amazing phenomena in cultural history. Between the middle of the seventh and the middle of the eighth centuries, the Bedouin tribes of the Arabian Peninsula had risen to the role of masters over most of the former ancient world. In short order they then had lifted themselves from the level of nomads to that of the brilliant heirs of the ancient cultures. Like the

Indo-European (or "German") tribes to the west, they had shared in the demolition of Rome, which in its turn had held together much of the ancient world. Yet unlike the Germans, they had conducted their worldwide conquests from a solid Oriental base containing the ancient cultural centers. That difference proved a weighty element in shaping their historic role.

Three factors chiefly destined them for their brilliant cultural career. One was the native vigor of a people largely unspoiled by civilization, avid to assimilate the cultural legacy that was spread before their eyes in the freshly conquered lands. A social group long excluded from education usually develops extraordinary energies when faced with a chance to close the gap. In this instance the "cultural outsiders" came mostly from a level equivalent to prehistoric society; through their conquests, the centers of practically all ancient civilizations had fallen into their hands. Their capacity for learning seems to have been proportionate to this cultural gap.

The second factor was the Koran. Muslim history and culture (including Islamic science) were decisively inspired by the teachings of this holy book. With its strict monotheism, the Koran had welded its followers into armies fighting for a divine mission, into communities whose daily life was regulated in every aspect by religious laws, and into a civilization historically distinguished by its basic open-mindedness toward alien cultures, stemming from the complete assurance of its own religious beliefs — its unflinching submission to the one God and the word taught by Muhammad, Allah's Prophet. It was this elementary inner assurance — at the peak of its power, at any rate — that gave the Muslim community its extraordinary flexibility in confronting alien civilizations and their legacies.

But the most vital factor in the evolution of Muslim science was the cosmopolitan character of the culture the Arabs came to create. Spanning the ancient world from the Ganges to the Atlantic, Islamic civilization united within its scope the cultural traditions of India, Persia, and Mesopotamia, of Egypt, of large parts of Byzantium, and of the Greco-Roman legacy developed by the Roman Empire in the Mediterranean West. The Arabs proved to be masters in weaving all these different strands into a new cultural fabric. The new civilization was held together by their common language, their common faith and common way of life, yet it was general enough, at its height, to tolerate the free exchange of all these original diversities. Towns from India, all across the Near East and North Africa, to the coast of Portugal have preserved the imprint of that heterogeneous culture into the present.

Within this enormous arc the scientific legacies of nearly all ancient civilizations were merging into the world of Islam. The trade routes of a pulsating commercial life aided the flow of ideas and knowledge. A Muslim could study, from records preserved on his own soil, the astronomies of India, Babylon, and Egypt; Indian and Persian mathematics; the philosophical concepts of the Greeks; the medicine, geography, astronomy, and mathematics of the Hellenistic age; the botanical, pharmacological, zoological, geological, and geographic lore amassed by the ancient world as a whole. With the lone exception of the Far East, the entire evolution of science from its earliest beginnings in the Nile Valley, between the Euphrates and the Tigris, or along the Indus banks had, in fact, occurred in what was now the Muslim world. Islamic science began with the natural fusion of all these precious legacies. And through trade, Islam even served as an important link between the rich Chinese legacy of technology and science and the needs of the West.

Medieval scholars crossing the Pyrenees found the quintessence of all preceding science distilled by the theorists and practitioners of Islam. Historically, by entering the area of Islamic civilization they had indeed entered the whole vast vibrant world of antiquity as well.

* * *

Islam — in contrast to the still largely rural, feudal, and severely ascetic civilization in which these scholars were at home — was urban, commercial, sophisticated, exotic, and cosmopolitan. It was also decidedly world-oriented; according to the Koran, the Muslim faith was to be practiced in the social ethics of daily life.

Islamic science showed the imprint of all these cultural traits: it bore the stamp of the emphatic social consciousness of the Muslim religion; of the pragmatic attitude of the commercial life; of the urban character of Muslim society; of the rich diversity of the many different cultures that had gone into the making of Islam.

Where Western science came to unite itself with the strong technological bent characteristic of the northern-European people, Islamic science was primarily colored by social and commercial interests. Where Western science on the other hand evolved from the beginning within a tight framework of theoretical thought, Islamic science was essentially marked by a happy-go-lucky diversity of philosophical views, reflecting the crazy-quilt pattern of its various component cultures.

Islamic science (despite the brilliant thought of philosophers like Avicenna or Averroës, even despite Aristotle's enormous impact) seemed to lack the need for a consistent intellectual framework — a compelling need the West inherited from traditional Medieval philosophy. Science in the Muslim world was inspired by the pleasurable observation of nature's diversity and the uses of its bounty for the enhancement of life. It had little concern for establishing the mastery of the mind over nature through tight philosophical systems, or for proving human power through the relentless technological transformation of the natural environment. In its most vital essence, Islamic science was the pragmatic product of a religious culture that looked at the earth as a garden, not as a testing ground for the power of the human race.

Over and beyond all specifics, it was this delight in the variety of the natural detail and its uses for society that the Middle Ages learned from Islam. Under the impact of this encounter the West took the step toward the cultivation of specialized sciences, out of the original philosophical core. Every single specialized science in the West owes its origins to the Islamic impulse — or at least its direction from that time on. It was from Islam that the Middle Ages learned to look on nature as an infinitely varied reality, not as a philosophical idea. Till then, the West had considered science as a type of philosophical thought (so that William of Conches quite logically called it *philosophia*). The encounter with Islam changed that conception to the modern one, of a diversified body of specialized knowledge. *Philosophia* evolved into *scientia*.

* * *

It was in the Muslim world that the West first met with a highly developed system of medical care. Here, Europeans for the first time saw independently functioning hospitals, in contrast to the infirmaries of the monasteries they had known at home. The institution had been founded by Harun al-Rashid in his capital, Baghdad. By the high Middle Ages there were more than thirty hospitals functioning throughout the Arab world, complete with women's and maternity wards arranged around a series of patios, each with a splashing fountain or date palm in the center, with surgeries, dispensaries, and sometimes medical libraries and schools.

Western observers found a medical profession from which, ever since the tenth century, quacks had been screened out through a careful system of examinations. At the street corners of Muslim cities they could see the first apothecary shops the world has known, dispensing Oriental spices and herbs, preparing medications according to prescriptions — already equipped with the many-colored phials, jars, bottles, mortars and pestles arranged on high shelves that lend such shops their exotic and antiquated atmosphere even today.

Behind his counter the Muslim pharmacist already practiced his art with the help of a pharmacopoeia, a book describing herbs and medical preparations compiled on the authority of some of the greatest scientists in Islam — the same basic reference book the pharmacist, with the additions and modifications of the ages, still uses today.* The kind of social concern, inspired by the Koran, that was behind such emphatic care for the sick benefited even the poor Arab villages through a kind of rural health service, and Muslim physicians had adopted the practice of visiting jails.

Islamic doctors, especially those of Persian origin, had made important contributions to the development of surgery. The famous al-Razi, chief physician at the hospital of Baghdad around 900, remembered as a great and gifted clinician, is credited with a long list of ingenious devices — like the idea of introducing a few threads or horsehairs beneath the skin to produce an issue (the "seton"). His portrait, together with that of another outstanding figure in Islam medicine, Avicenna — also a Persian by birth — can be seen in the great hall of the School of Medicine at the University of Paris. Al-Razi (whose name the European scholars latin-

---

* Muslim pharmacists as well as physicians had to pass an examination, ever since al-Mamun, caliph of Baghdad in the early ninth century and one of the most energetic patrons of science, had introduced that requirement.

Examples of cauterization, used for cancer and to open abscesses, from the First Turkish Medical Manuscript of 1465, show techniques introduced by Islamic physicians.

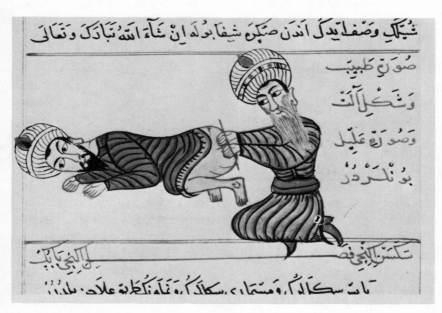

The extirpation of hemorrhoids in this amusing illumination, from the same manuscript, illustrates the continuity between Greek and Islamic medical practice.

ized to Rhazes) wrote an encyclopedic compendium of medicine — Greek, Hellenistic, Hindu, Persian, as well as Arabic — entitled *al-Hawi*. Together with his *Liber Almansoris* and Avicenna's *Canon*, it was used as a textbook in European medical schools into the early modern age.

Works like these, which the Christian scholars found in the Spanish libraries and eventually translated into Latin (the *al-Hawi* was, however, translated by a Sicilian Jew), opened the door to the entire history of medicine. The Islamic compilers had drawn on the accumulated medical knowledge of almost a thousand and five hundred years past, from Hippocrates in the fifth century B.C. through Galen in the second century of our era, to Islam's cumulative contributions, plus whatever earlier medical know-how had entered into this powerful stream. From their own native countries the Medieval scholars knew only the crudest kind of empirical medicine shot through with all manner of magic. What Islam had to offer them now was not only a spate of enlightening digests of this whole, long, rich evolution but an intelligent discussion of all its essential features, screened and refined through Islam's own intensive experience.

Beyond countless individual data about diseases and treatments, what Islam had learned from Greek medicine (and what Europe now came to learn from Islam) may primarily be described as an attitude toward the phenomenon of illness. Instead of regarding it as an incomprehensible disaster, the work of evil spirits, as primitive cultures might do, the Greeks had recognized that disease is a natural process, part and parcel of the physiological patterns of the human body, often brought on or exacerbated by the inevitable stresses of life. It was logical, therefore, that the Greeks taught the physician to rely to a substantial extent on the patient as the natural source and subject of illness, both for diagnosis and cure, by asking the right questions, by evaluating answers and symptoms, by encouraging the sick body as much as possible to regenerate itself. Because of this wholesome emphasis on human nature and its recuperative powers (an emphasis typically Greek in that it reflected faith in both nature and reason), Greek medical literature is rich in the careful observation of symptoms, the exposition of clinical methods, the description of "natural" cures. Greek doctors had studied the human body both in its healthy and pathological states — an evident sideline to art's glorification of the body — amassing centuries of experience in the kinds of diet or exercise or herb that would prove effective in restoring healthy balance.

By about 900 this entire tradition had been assimilated by Islam. The following two centuries witnessed a great flowering, something like a golden age of Islamic medicine. The hospitals permitted the study of a great variety of special diseases. Al-Razi, from his vantage point at the head of the hospital at Baghdad, systematically observed diseases like measles, smallpox, bladder stones, and kidney stones, and summarized his findings in case histories or monograph studies, leaving a deeper understanding of the nature of illness — including new categories of diseases — to the future. The very vastness of the Islamic Empire had a stimulating effect: It encouraged the observation of a wide variety of drugs, facilitated the exchange of Greek medical literature among far-flung intellectual centers — and, on the other hand, netted a crop of treatises on dietary rules to be observed during travel in different climates.

Western Europe had already felt the influence of this rich medical culture, a century before the translating activity reached its peak, when a highly important figure in Medieval science called Constantine the African had rendered a respectable portion of Arabic medical literature into Latin. His translations include some of al-Razi's works as well as the studies of a brilliant Jewish physician from Egypt who came to be known in Medieval Europe as Isaac the Jew. Constantine's translations had been a great help for the budding medical school at Salerno. Medieval medicine, fed by these influences, experienced an early flourishing in Southern Italy and Sicily. Still, the real contact with Arabic medical culture was made by the twelfth-century translators in Spain: it was then and there that the dikes were opened and the stored-up experience of the ages began to pour in upon the Medieval West.

From Islam, the West also learned the concepts and methods of the alchemists, complete with their laboratory equipment and techniques — a quaint half-mystical, half-experimental tradition on which Medieval Europe seized with predictable gusto, yet which in the end led to the rise of modern scientific chemistry.

Islamic science grew out of the Muslims' love for the world, their passion for reproducing its exact features. This penchant caused them to leave a host of measuring instruments and observational data. A number of astronomical observatories were scattered throughout the Arab world, ever since Caliph al-Mamun had established the first ones at Damascus and Baghdad. The Arabs had compiled astronomical tables, records of a systematic observation of the stars. They had developed — or improved upon — such strategically important instruments as the astrolabe, the sundial, the armillary sphere. They had produced careful catalogues of herbs and plants based on original Greek and Hellenistic studies; instruments to measure optical refraction; amazingly accurate calculations for measuring the length of a degree. Some inventions, which had their practical applications for every day, had developed from these empirical studies. The need for minute business calculations was met by the introduction of a highly simplified numerical system (based on the zero) that proved of enormous help for the businessman — and became one of the most momentous contributions of Arabic science. The intensive literary activity, both in science and other fields, drove home the scarcity of the ancient writing materials, causing the invention and manufacture of writing paper for common and luxury use. And the emphasis on medical care produced a whole basic set of surgical instruments.

* * *

Much of this was so closely woven into the fabric of everyday life that a foreigner could have formed an impression of Islam's scientific culture simply by strolling through a Spanish town or inspecting a hospital or a former Muslim business establishment. By contrast, the libraries opened the view into the world of Arabic theory and thought. But there were staggering obstacles to overcome, quite beyond the mastery of the language.

Obviously, the Medieval scholars lacked any conception of specialized science, for the simple reason that the West had not yet advanced to this point. And so the translators had to acquire a grasp of the elemental structure and methods of science, to learn what the scientific approach was all about, at times too to penetrate complex mathematical or astronomical problems — even as they stumbled through the Arabic characters, trying to piece together their literal meaning, sentence by sentence and word for

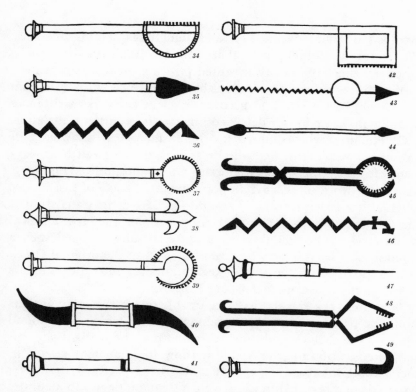

Medical instruments, illustrating a book by the eleventh century Islamic physician al Bukasis (although in a somewhat later edition), include scalpels, dental, obstetric, and other surgical instruments.

word. They had to cope with the dual bewilderment of language and content. What is surprising is not that the results were quite often defective, but that they managed to make the formidable contribution they did.

As a matter of fact, errors abounded. Historians tend to think of the work of the translators as though they had lifted the Islamic legacy lock, stock, and barrel from the Arabic texts and neatly inserted it into the context of Western thought. In reality the translators' work was haphazard and often slipshod by any rigorous standards. It was without overall direction, the choice of texts based more on the interests of the moment than on any criteria of scrupulous inclusiveness — betraying an erratic tendency by some of the translators to work on several texts at a time — which involved frequent duplication and sometimes serious mistakes. The results left crucial gaps that were not filled till the Renaissance. By then, the newly invented printing press was kept busy turning out improved translations by Renaissance humanists from the Greek, besides freshly discovered scientific works written in Latin, to make up for the defects of the twelfth-century translators.

Of Ptolemy, who summed up the cosmological knowledge of antiquity, working in its twilight as he did, the twelfth century translated the *Almagest* as well as his *Optics* but omitted his *Geography* — a work of decisive importance for the age of discoveries when at last it was translated around 1410. Even the translation of the *Almagest,* with its critical influence on the astronomical world-view of the Middle Ages and Renaissance, proved such a careless job that it had to be retranslated from the original Greek during the fifteenth century by Regiomontanus and other students of the German astronomer Georg Peurbach.

Sometimes the translators were unable to tell the original texts from the package of Arabic commentaries in which they found them on the shelves, and thus treated both as the work of one person, often attributing the Arabic writings to an ancient author. In many cases manuscript pages that had been partly written over with a second text (palimpsests were a frequent device in a culture where writing materials were still at a premium) were read as though both texts had been written by the same author — baffling though the results must have sounded even to the translators themselves.

Translating techniques were still primitive. At times, the translators would simply use the method of "linear translation," by which a sentence is rendered word for word and line for line rather than from its basic meaning. Often what was purported to be a translation was merely a paraphrase or a random collection of excerpts, freely interspersed with the translator's own comments. Subsequent translators have in some cases explicitly shown that the twelfth-century version was grossly inaccurate. The anonymity of many translators (or the often highly casual identification of translators' names and works) has resulted in a host of manuscripts in which translated texts and original treatises — sometimes by various authors — are jumbled together in a confusion virtually beyond hope.

About a century after the bulk of the translations had been completed, Roger Bacon sneered broadly at the translators who, he said, "had the arrogance to translate innumerable writings [although] they knew neither the sciences nor the languages, not even Latin, and in many places they inserted words of their own mother tongues." Though Bacon had every reason to use these harsh words, his judgment seems a bit like that of the son who feels superior to the less sophisticated father, forgetting in his own arrogance how much he really owes to the paternal spadework. The truth is that Bacon's vision of a scientific method and its revolutionary possibilities for a future age would have been inconceivable without the work of these often clumsy pioneers.

A little over a generation before the work of the translators was reaching its peak, the masters of Chartres had made an effort to put together a systematic library of ancient science. It had been a meager shelf. Besides Plato's *Timaeus* in Chalcidius' incomplete version (and what information the *Timaeus* netted of earlier Greek thought), there were some fragments from Pliny's *Natural History;* some portions from the writings of Macrobius; some of Boethius' mathematical exercises; a hodgepodge of classical science and popular superstition compiled by a well-meaning archbishop in the late twilight of Rome, the *Etymologies* of Isidore of Seville; and a few more odds and ends. Though the masters of Chartres managed to expand this paltry library to some extent, mainly by reinterpreting the texts, this was essentially the whole legacy of science that had survived the collapse of antiquity. Toward the end of the thirteenth century, when Bacon was formulating his system, he had what amounted to the whole evolution of classical science at his fingertips, from Aristotle through Euclid and Archimedes to Galen and Ptolemy, a solid five hundred years of ancient science at its brilliant and specialized peak — plus, of course, the immense body of Islam's original contributions.

The gap had been filled by the twelfth- and early thirteenth-century translations from the Arabic (even though a few direct renderings from the Greek original began to round out the picture already in Bacon's time). The efforts of two generations had given Medieval science a historic basis, placed it in the mainstream of history, transformed an isolated body of philosophical speculation into an integral part of a long chain of consistent thought.

This was a momentous achievement, transcending the limits of science, even transcending the flaws and inaccuracies of any individual text. No doubt Roger Bacon, like most scholars, turning up his nose at the translators' errors, overestimated the importance of textual accuracy for the life of the mind or underestimated the freewheeling life of ideas, the ways in which a body of writings may profoundly influence a culture through implications, associations, hidden impulses, subtle stimuli, quite beyond the literal meaning of the word.

The Arabic translations raised before Medieval eyes the exciting image of a mature civilization in which science happened to play a major role. Beyond that civilization, like a background panorama of successive mountain chains, they conjured up the intellectual landscape of Greece. Quite correctly, the translated texts showed the Greek mind as having been crucially concerned with scientific problems, rather than reposing in a vacuum of pure abstract thought.* The resurgent image of the classical world — and, with it, the first stirrings of historical consciousness — came to the European mind very largely in the shape of science.

Even those flaws and inaccuracies — certainly a grave matter, especially in scientific texts — had their salutary effects on European culture in the long run. By gradually forcing the scholars of following centuries into careful textual comparisons and emendations, these very errors helped to stimulate the sense of philological accuracy among future humanist generations. If Europe managed to lift itself from the status of a half-barbaric society to a center of pulsating creative culture, those crude translations of scientific texts played a vital role: they helped in piecing together the shattered image of the classical world and reweaving the sense of cultural continuity.

* * *

* It seems significant that Aristotle electrified Medieval Europe first and foremost as a scientist, and that the bulk of his philosophical writings — only very few of which had been known before — were accepted as a kind of methodological framework for his science, and a straight appreciation of Aristotle the philosopher developed as something of an afterthought. Just as the first of Plato's dialogues widely studied in Europe had been his breviary of science, the *Timaeus*.

Scholars avid for discovering Arabic manuscripts had begun to cross over into Spain since the end of the tenth century, when Gerbert of Rheims, who later became Pope Sylvester II, went to Catalonia to study Arabic mathematics and astronomy. At first, the scholars came in a trickle; by the early twelfth century they appeared in droves; by the second half of the century they started to act somewhat like an effective (though of course unorganized) team, eagerly working away at the manuscripts till the most important were translated. Their ranks were swelling apace with the consolidation of Christian control over Spain, the gradual if still unsteady pressing back of the Muslim forces.

The volume of the translated works — and the proportion of Islamic science thus made available to the West — roughly followed this rhythm. It reached its peak in the work of Gerard of Cremona, a giant of literary productivity who translated more than seventy works from the Arabic between his arrival at Toledo in 1160 and his death twenty-seven years later. By the time Gerard came to Spain, enough had already been translated for him to spot some of the major gaps and decide to fill them; above all, Ptolemy's *Almagest* and the core of Aristotle's scientific writings. Both are included among his achievements.

The monumental scope of Gerard's contribution can be gauged from the list of titles of his translated works, which was compiled by some assistants or "students" and is still available for our reference. The goal he seems to have set for himself was nothing less than to cover the whole breadth and depth of Greco-Islamic science. If the *Almagest* represented something like the balance sheet of ancient astronomy, Gerard's choice of his further texts shows that he sought to extract the essentials of every important field from the Arabic manuscripts: in medicine, two of the chief Islamic compendia (Avicenna's *Canon* and al-Razi's *Liber Almansoris*); in optics, two fundamental works by al-Kindi; a work on acoustics (al-Farabi's commentary on Aristotle's *Liber de naturali auditu*); a study on chemical substances by the prolific al-Razi; as well as writings in the fields of geology, physics, mathematics, mechanics — including major portions of Euclid's *Elements* and Archimedes' *On the Measurement of the Sphere*. When one adds Aristotle's major scientific works to this list — his *Physics*, his *On Heaven and Earth*, his *Generation and Corruption*, and the first three books of his *Meteorology* — the full magnitude of Gerard's ambition stands revealed. By his single-handed effort he wanted to move Medieval science in one giant push from the speculative phase to the high level of specialization it had attained between Greece and Islam.

The most remarkable fact about this is that Gerard succeeded. Selecting his texts with a keen eye for what was most relevant and intrinsically best, Gerard of Cremona, forerunner of the great Italian Renaissance humanists, in his lone way did more than any successor in closing the formidable gap. Others, English or Scottish, German or Flemish, Spanish Christian or Spanish Jew, may have refined and supplemented his work or, later, corrected his errors. But it was Gerard's translations that were destined to influence European thought most lastingly and deeply.

His medical texts remained the basis of medical training in Europe for the following five centuries. The optical studies by al-Kindi, which he translated and which embodied substantial advances over ancient optical science, became the foundation for a theory of visual perception that culminated in the formulation of the laws of perspective during the Renaissance. Euclid's *Elements* — probably the most widely distributed book in Western civilization next to the Bible — eventually went through more than fifteen hundred printed editions. And Gerard's vigorous grip on Aristotle's work proved crucial in introducing the complete body of the Greek philosopher's extant work to the European West. If the Scientific Revolution was to start with a radical revision of Ptolemaic astronomy while operating within the framework of Ptolemy's thought; if the pioneers of modern science were to employ the tools of Euclidian mathematics, refined by the work of the last three hundred years; if they eventually managed to shatter the Aristotelian cosmos largely by using Aristotle's own scientific logic and methods, it was Gerard of Cremona more than anyone else who had supplied them the implements.

★　★　★

And yet, the translations were very much a collective venture. Their bulk was not confined to one man nor one time nor, for that matter, to any single country. In fact, their setting included almost the entire span of Islamic civilization circling the Mediterranean. Wherever Europeans had any intensive contacts with Islam, Arabic science was reaching the West through the work of individuals with a feeling for the future.

In Syria, during the early thirteenth century, Philip of Tripoli translated the *Secret of Secrets* — a famous Arab book that gave Roger Bacon the idea for his surreptitious method in discovering nature's secrets, and made an enormous impact on a mystical current in Medieval scientific thought. In North Africa, there was not only the work of Constantine the African in the early eleventh century; around 1200, Leonardo of Pisa wrote his epochal introduction to the Islamic system of algebra — a system originally developed by Hindus, Persians, and Arabs — which brought the famed Arabic numerals to the West.

Yet the most important meeting ground after Spain was Sicily. The island had been ruled by the Arabs during the tenth and eleventh centuries and was still the scene of profound Arabic influences in the twelfth and thirteenth, in particular under two inspired rulers, the Norman King Roger II and Frederick II, the famous and colorful emperor. Both had their minds open toward the Arab world and the promise of science. Islamic geography, astronomy, zoology, optics were cultivated at the Palermo court. Al-Edrisi, a North African Muslim, wrote his *Geography* for King Roger; and Frederick conducted a lively correspondence in Arabic with Muslim scholars on various scientific problems. Sicily and the Italian South, still filled with Arabic traces today, were a major gateway to Islamic civilization.

Frederick had grown up as an orphan in the Sicilian capital, naturally absorbing Arabic, which was still the language of the streets. As an adult he presented the phenomenon of a Christian ruler fully versed in Islamic culture — a Christian ruler, however, who always maintained a critical and bitingly sarcastic detachment toward the whole Christian community, undoubtedly tinged by his childhood amongst the Arab people of the Palermo streets.

The strong scientific interests of his manhood years were a fruit of that background. In the midst of a stormy political life he found time to write an enchanting book on his favorite hobby, the falcon hunt, which became a model for later zoological studies into the Renaissance. He befriended Leonardo of Pisa and absorbed his revolutionary mathematical theories. Most important, he attracted the great Michael Scot to his court, chiefly as a discussion partner on zoological and astrological questions, two of the fields in which the brilliant Scotsman excelled. Frederick even introduced the Islamic requirement of a fixed term of studies for doctors in his Sicilian realm.*

By the time Scot appeared at the court of Palermo, around 1227, he already had a respectable list of translations and original scientific works to his name.† Ten years earlier he had started out as one of the translators in the Toledo group. Belonging to a younger generation than Gerard of Cremona, he had certainly profited from the work of these pioneers. His heightened scientific understanding endowed his writings with an extraordinary influence, second only to Gerard's. By translating an important Arabic commentary on Aristotle's cosmology, al-Bitrogi's *On the Sphere*, he managed to free the Aristotelian astronomy from the overshadowing influence of Ptolemy's *Almagest* and present the West with a kind of "pure" Aristotelian system. He thereby unleashed a controversy between the advocates of Aristotle's and Ptolemy's respective cosmological systems that was to worry Western science until the eve of the Scientific Revolution. It was a fertile controversy: since the mathematical problems needed for an intelligent choice were extremely complex, the two systems acted as an effective incentive for Western mathematical thought.

What is more, Scot had translated the Spanish-Arab Averroës' commentaries on Aristotle's philosophy — and thereby not only

---

* The emperor proved his admiration for the more sophisticated Islamic institutions by introducing a number of Arabic features into his modernized Sicilian State — for example, the regular collecting of customs duties as a source of government revenue. Not only that institution but also its name in several European languages goes back to Frederick's initiative: *diwan* as the "seat" of the Arab customs collector — or, generally, the department of revenue — became *dogana, doana, douane.*

† Lynn Thorndike (*Michael Scot*, London, 1965, pages 32 and following) has argued for a somewhat different chronology, with Frederick asking Scot to enter his service in 1220. Thorndike believes all of Scot's writings of undisputed authenticity (other than his translations) were done at the emperor's request. In fact, Scot's three-part opus — *Liber introductorius, Liber particularis,* and *Liber physiognomiae* — is addressed to Frederick II and contains answers to questions by the emperor. But its date is uncertain and the answers may have been inserted into an already completed text, so Scot could easily have begun writing before his Palermo days.

---

contributed to the profound impact of Aristotle's logic on Western thought but, even more significantly, to the rise of the "Averroist" movement. Though somewhat vague in its philosophical substance, Averroism served as a militant rallying point for a radical brand of scientific rationalism for the next two or three centuries, sending shock waves through the University of Paris in Thomas Aquinas' time.

Scot was also the first to bring Aristotle's zoology to the West, replacing the fairly primitive Medieval notions about the animal world with the Greek's broad and trenchant classifications. The work he translated within a few years — Aristotle's On Animals, of which he completed the standard Latin version around 1220 — comprised three of the Greek's major writings (The History, the Parts, and the Generation of Animals), or nineteen books in the Arab version. Once more the effects were momentous: taking Scot's translation as a basis for further original research, Albertus Magnus was able to launch zoological science in the West, essentially by applying the Aristotelian categories to the fauna of central and northern Europe.

At the court of Palermo the emperor was able to share his interests with Scot. Since Michael Scot also wrote on (and practiced) astrology, Frederick named him his official court astrologer and so could consult with him freely on the secrets of the heavens. Later, Dante was to consign the Scotsman, for this line of interest, to the Inferno as a false prophet, "practiced in every slight of magic wile." Astronomy was still mixed up with the foretelling of the future, science with mysticism, when these two great Medieval figures — the magician-scientist and the emperor reputed to be in the pay of hell — were conducting their discourses, perchance in the deep of night, casting their shadows against the walls of the imperial palace in the Sicilian capital.

* * *

All along the common cultural frontier then, Arabic science was seeping through to the European West, together with Muslim everyday customs and governmental institutions, a host of Arabic terms, the colorful decorative patterns and elegant features of Arabic architecture, the whole easygoing luxurious style of Muslim upper-class life. Coming of age, Medieval Europe was beginning to absorb the sophisticated ways of a more advanced civilization — even though the new ways were to be modified by the cultural traditions, the intellectual attitudes, and, sometimes, the native superstitions of the Medieval West.

Generations of scholars sat in front of the Arabic texts deciphering the exotic symbols, in reading rooms from Syria to Portugal, though more than anywhere else in the libraries of Spain. Outside, a fierce Spanish sun might beat down upon patio or cloister. Inside the halls, built for coolness, nothing was heard but the rustling of manuscripts, the quiet music of scholarly study. The great pilgrimage to Islam had climaxed in a period of concentration, in the silent service of the word.

What final essence were the manuscripts yielding to these long and devoted labors? Just as in medicine, the Arabs had everywhere done more than merely transmit Greek and Hellenistic science. They had condensed the substance of classical science in encyclopedic and often penetrating digests and then added their own comments, usually betraying their characteristic bent toward the empirical, the specific, the emphatically concrete. This was as true for the more theoretical sciences like mathematics, physics, or astronomy as it was for the study of diseases.

Everywhere, too, the Persian influence was strong, mostly because of a curious twist of historic circumstances. Back in the sixth century, when the Byzantine heirs of Hellenistic culture had stifled the Greek tradition of free inquiry — the closing of the School of Athens by Emperor Justinian in 529 had been part of that ominous trend — scholars and scientists, especially of the Nestorian sect, had migrated toward the Persian East. They found a particularly warm welcome at places like Gundeshapur, Edessa, or Antioch. It was from these centers (and, somewhat less, from Alexandria) that the Greco-Hellenistic legacy had bloomed forth in Islam. The last embers of ancient Greece were feeding the flame of Arabic culture. The evolution of science follows its own strange itinerary, mapped by the vicissitudes of intellectual freedom.

Islamic observatories had sprung up on Persian soil, like the one at al-Rayyi near the modern city of Tehran headed by Omar Khayyám, the great Persian scientist and poet; or in neighboring areas, like the one at Samarkand, which the visitor may still inspect, with its marvelous equipment, including its astrolabe, an ancient precision instrument for measuring angles in the changing position of the stars and thereby charting their courses. It was through meticulous observations of this type and their recording in astronomical tables (in the clever shorthand of the Arabic numerals) that Islam expanded our knowledge of the heavens. Our night sky is sprinkled with stars bearing Arab names — like Mizar, Alcor, Aldebaran, or Betelgeuse — heavenly witnesses to the scrutiny of Islamic observers of a thousand years ago.* Yet such observations were far from haphazard: Islamic astronomers were working within the tightly explicit astronomical systems of Aristotle and Ptolemy. To have transmitted the skies of the Greeks and enriched them with their own watchful observations represents the full Muslim contribution to modern astronomy.

Similarly, their chief contribution in physics was in the comparatively concrete branches of optics and mechanics, to which the Arabs added some minor technological facts. In the more theoretical aspects of physics, on the other hand, the Arabs did little more than transmit Aristotle's great explanatory system, leaving it to western Europe to make the crucial advances that were to surpass the Greeks'.

The general trend was, in short, for an application of Greek methods and concepts to original observation — and in a few cases the distilling of fresh theoretical conclusions from the tangible evidence that had been amassed. "Seeing" and the precise identification of what one saw, as well as the optical laws that are ruling our visual observation, such was the thrust of Islam's approach, the prime source of its original creativity. What a Medieval Europe, its vision trained on purely spiritual things, learned from the Orient was, in the final analysis, the proper use of the eye. One senses a legacy of the keen-eyed nomadic tribes of the ancient East — Iranians, Arab Bedouins — just as the Germanic people contributed their native technical ingenuity to the growth of science, and Medieval culture its genius for abstract thought.

*   *   *

* Alcor and Mizar form a visible double star in the center of the handle of the Big Dipper. It is said that Muhammad tested warriors by seeing if they could spot Alcor and Mizar with the naked eye.

The two magic names in Islamic optics, the translators discovered, were al-Kindi and Alhazen, whose Arab name was really ibn-al-Haitham.* Al-Kindi, translated by Gerard of Cremona, had refined Euclid's *Optics* during the ninth century. Alhazen, working in Cairo around 1000, had continued in the broad tradition of Greek optical studies, extending from Aristotle and Euclid to Ptolemy. Even more than al-Kindi, Averroës, or Avicenna, he became the prime source of optical knowledge for the European Middle Ages and Renaissance. Men like Roger Bacon, Leonardo da Vinci, Johannes Kepler were inspired by his insights and influenced by his methodological approach. If Islam was teaching Medieval Europe to see, Alhazen taught the most incisive lesson in visual precision; no wonder the scientist-artist Leonardo de Vinci was his particular admirer.

By applying sophisticated geometrical methods (plus exact measurements) to optical research, Alhazen had carried Greek studies on the reflection and refraction of light to a point that remained virtually unsurpassed — or, in other words, accepted as scientifically accurate — till the advent of modern optics. For instance, where Euclid and Ptolemy had already identified rays as the transmitters of light, Alhazen established that the rays originate with the luminous object, not, as the Greeks had assumed, with the eye. He also extended the study of reflection from plane surfaces to more complex geometrical bodies, such as concave and parabolic mirrors; and related the laws of refraction to the respective transparency, or density, of the medium by which the light is deflected from its course (including the atmosphere). In all this, he displayed a masterly combination of instinctive physical understanding (especially with regard to the laws of motion), empirical curiosity, keen geometrical analysis, and a remarkable inventive ingenuity in the use of mechanical apparatus, like the construction of steel refractors with the help of a kind of lathe.

Translated in part by Gerard of Cremona and his contempo-

* Overwhelmed by a flood of Arab proper names, besides all the outlandish subject matter, the Europeans displayed a knack for Latinizing or otherwise westernizing these alien sounds — even as they merrily transformed the great Persian physician, physicist, and philosopher ibn-Sina into "Avicenna," making him sound like a native of Italy. Al-Kindi's Arab name was really Abu Yusuf Yaqub ibn-Ishaq ibn-al-Sabbah al-Kindi, and one is not a bit surprised that the translator preferred to simplify that.

raries, more accurately and fully during the sixteenth century, Alhazen stands as a brilliant example of Islam's creative role in the evolution of science — the way it built organically on the Greco-Hellenistic foundations, and in turn influenced western-European thought, till the age of Newton, through its genius for perceiving the concrete dimension of the world.

* * *

That Arab science should have excelled in optics seems the manifestation of a deep-rooted cultural trait. One wonders how a science as abstract as mathematics could have been part of this inclination toward the visually concrete. Yet it was.

Once again, both Islam's historical role and cultural temperament combined to color a science. There had been two broad streams of mathematical thought in the ancient world, two great traditions with their different types of mentality and approach, both of which were inherited by Islam. The Greek, with its natural emphasis on form and visual abstraction, had produced an essentially geometrical school. (Characteristically, even Euclid's great synthesis of Greek mathematical thought — George Sarton has called it "a monument that is as marvellous in its symmetry, inner beauty, and clearness as the Parthenon" — dealt with algebraic problems in geometric terms.) The second strong current, derived from Babylon and India, had centered on calculations and numerical symbols — perhaps under an original commercial impetus — and was therefore conducive to an arithmetical imagination.

The two currents had met, somewhat marginally, at several junctures in ancient history. Euclid himself had undoubtedly benefited from the encounter of Greek and Oriental science in Hellenistic Alexandria. A good eight hundred years later, by the sixth century A.D., those Nestorian Christians fleeing from Byzantine persecution of free thought had again taken the Greek tradition — mathematics as well as astronomy — and brought it along to their new homes in the Persian East. Finally, with the Muslim conquests, a new cultural metropolis arose in the capital of the early caliphs at Baghdad, located on the Tigris River, looking toward the east, wide open to Persian and Indian influences and to all the wisdom these cultures had stored up over the centuries.

From Caliph al-Mansur to Harun al-Rashid, to his son, that great patron of science al-Mamun, Baghdad was really the intellectual center of the Muslim world, a new melting pot for the Indian and Persian traditions and the Greek and Babylonian legacies that they encompassed. The blending of the two great currents in mathematics was part of that fusion. Islam had not only perfected it but handed it on to the future, as the basic mathematical equipment with which scientists are still operating today.

What we said about Alhazen's optics may illustrate how the geometrical legacy of the Greeks was stretched to solve fairly tangible problems in physics. Yet Islam's most durable contribution — that of another commercial culture, after all — was made in arithmetical mathematics. So powerful was the imprint Islam has left in this field that our numerical system is still called "Arabic," overshadowing its actual Indian and, in all likelihood, ultimately Babylonian origins.

* * *

As a matter of fact, whether Babylonian arithmetic had influenced Indian or — somewhat less likely — the two cultures had invented their numerical systems independently of each other is not entirely clear. What is certain is that both had progressed to a highly simplified system of "local" or "positional" numbers — that is, numerals that through their relative positions tell us what units they represent. In fact, both had devised a symbol for the idea of the zero, which the Indians called *sunya*, the "void." (The Arabs translated this concept as *sifr*, from which our word "cipher" derives.)

Most likely the culture of Babylon — whose bustling commercial life is reflected in its great law book known as *Hammurabi's Code* — had felt a compelling need for making quick, simple notations of prices, of interest and exchange rates, growing out of its intense business dealings. The Indians may have encountered this system during their lively trade contacts with Babylon in Hammurabi's time, around 1700 B.C.*

---

* In theory, of course, it could have been the other way around — Babylon learning from India — except that the Babylonians had inherited their mathematical foundations, along with most of their science, from the ancient culture that preceded them on Mesopotamian soil, the culture of Sumer, so it repre-

Hindu manuscript using the zero, indicated by heavy dots, as well as the 3, both in what we now call "Arab" numerals.

Whatever the timing of that historic exchange, India developed the Babylonian numeral system in several mathematically significant ways. With the metaphysical orientation inherent in the Hindu faith, she had refined arithmetical mathematics toward the theoretical side. Thus, the concept of the zero, which for the Babylonians had merely been a symbol denoting a "blank," was used in India for actual computation, sometimes involving high numbers and difficult problems. The Babylonian sexagesimal system (our method of counting seconds, minutes, hours still derives from that ancient practice) was transformed into the decimal system by Indian mathematicians, with the zero included as an integral part.

sented a very much older, eminently sophisticated legacy. But there were at least two later occasions in the history of the ancient East when such an exchange may have taken place as well: in the days of the Persian Empire, which had conquered Mesopotamia and maintained an active commerce with India till its conquest by Alexander the Great, or during the Hellenistic civilization, which sprang up in the vast stretches conquered by Alexander — witnessing a fusion between the Oriental cultures of unprecedented intensity (or, in other words, between the sixth and fourth centuries, or else at some time after about 300 B.C.).

The Hindus produced several gifted mathematicians who, in turn, produced several original mathematical works. A copy of one of these, known as the *Siddhanta*, showed up in Baghdad in Harun al-Rashid's time. This is how Islam fell heir to the arithmetical tradition — a set of signs and symbols, a mysterious (or magic) numerical code that embodied the sophisticated thinking and practical experience of some two and a half thousand years. Once more, then, the Muslims acted first of all as transmitters of a great historical legacy; but again they added their own contribution, and this time an absolutely momentous one.

* * *

The immediate credit goes largely to al-Khwarizmi, mathematician at the ninth-century court of al-Mamun in Baghdad — the man who fathered the terms "algorism" (derived from his name) and "algebra," as well as the concepts, at any rate for the West.*

The scholars of India had already traveled far into mathematical insights, in particular the arithmetical apprehension of infinity, or the use and meaning of infinitesimals. One suspects that the role of the Absolute (or Infinite) in Hindu philosophy had stimulated such speculations and that the computations based on the idea of zero, "the void," reflected these leanings.

Perhaps the best way to understand the Islamic contribution is precisely by examining that intrinsic visual emphasis. Using the kind of simplification that singles out the essential while ignoring certain nuances and shades, one may say that the Babylonians had developed a practical system of numbers inspired by intensely practical business needs; the Hindus had developed its mathematical potentialities in the direction of their native philosophical bent; the Muslims recognized the full visual implications of this system and perfected it into the marvelous computing device through which the West has been able to make its giant mathematical strides.

Al-Khwarizmi's achievement, expanded by other Muslim math-

---

* The Latin translation of one of his works opened with the words "*Algoritmi dicit*," meaning simply "al-Khwarizmi says," which, however illogically, established the term for calculating by means of ten figures, including the zero. The term "algebra" — from an Arabic word meaning the reducing and recombining of parts — was al-Khwarizmi's own title for his chief work.

ematicians (among whom the ubiquitous al-Razi, Alhazen, and al-Kindi were not missing), consisted in taking the Babylonian-Hindu numerals and converting them to an instantly workable code, one so simple that literally any child can handle it, so flexible that in the hands of the mathematician it becomes a vocabulary by which the most complex relations between the most astronomical quantities can be expressed. In short, al-Khwarizmi unlocked the doors to the everyday commercial, as well as advanced mathematical, use of the Hindu system, by demonstrating its basic operations and therewith revealing its instant — and unlimited — workability.

The Arabic numerals are, in fact, founded on an ultimately visual principle, and it is that which gives them their nearly unlimited usefulness. A promising trend in Medieval mathematics, culminating in the concept of universal mathematics, proceeded from the Pythagorean assumption that the entire universe follows a mathematical order, and that it must therefore be possible for one to capture its essence in mathematical terms. The idea has become perfectly familiar to us, but what about the mathematical expression? One need only to ask oneself how the motions of the stars, involving countless measurements of distances and angles, or the general laws of physical motion, or any other cosmic phenomenon, could have been expressed in the cumbersome terms of the Roman numerals to appreciate the simple eye-appealing quality of the Arabic system. It was a revolution on a par with the invention of the computer; one was able to reduce the cosmos to a system of ten elementary symbols, from zero to nine.

There are two ways in which the Arabic numerals operate by an immediate appeal to the eye. In contrast to the Roman numerals, each of the ten symbols on which they are based tells us at a glance its value; indeed, it parallels the principle of writing by means of the letters of the alphabet. What is more, their position in any multidigit figure will tell us with almost the same instantaneousness whether their value is to be read in the ones, tens, hundreds, thousands, and so on. The manifold advantages over more primitive systems spring at once to the eye; they rest in a highly developed, basically simple, and at the same time ingeniously flexible use of the visual symbol.

Only a single continuous tradition had advanced to this point. By contrast, what we call Roman numerals represent the crude level of number notations common to practically all primitive peoples in the ancient world, from the Egyptian to the Greek and indeed the Roman numeral systems. It is essentially a system based on counting with the fingers of one hand, sometimes extended to all ten fingers (and among some primitive tribes even the toes). Evidently the only relief in this kind of counting stems from the fact that the number of fingers or toes at the counter's disposal is as limited among primitive tribes as it was among the civilized Romans. It follows that some sort of symbol suggested itself in accordance with these natural limitations; that is, whenever a total of five, ten, or twenty was reached. All one could do in writing a number was jot down marks for the individual fingers or the signs for their totals — a two-level system of counting that lacks any differentiation other than the elementary limitations of the human anatomy.

The Roman sign for *five*, mirroring these origins, is almost certainly a symbol for the V-shape formed by the hand with the fingers held together and the thumb extended. The Roman *ten* was presumably formed by two of such fives, set apex to apex, one on top of the other. The lower Roman numerals — I, II, III, and, in the older form, IIII — represent nothing more than elementary notches (or "fingers"), the kind of marks someone might still jot on a blackboard for a quick vote count in a small-scale election. The whole system is nothing more than a primitive method of "bundling," by which the notches "pile up" or are bundled together, and the eye is forever losing itself in counting the individual marks. (At a somewhat more developed stage, higher units may be designated by particular symbols, usually the initial letters of the respective words — like the Roman C for *centum*, one hundred, or M for *mille*, one thousand. But once even that step was achieved, the bundling together — and the eye-confusing counting — had to start all over again.)

The superiority of the Arabic numerals to this system is analogous to the superiority of fully developed script, based on the alphabet, to the hieroglyphics of the ancient Egyptians. But where the hieroglyphics, or any of the other ancient pictograph scripts, work with an immediate visual appeal (so that the development of an alphabet involves a respectable feat in symbolic abstraction), the replacement of finger-counting and cumulative notches with instantly identifiable symbols, signaling their positional values by their place, represents something like the reverse of this process, a breakthrough to visual simplicity — although a considerable amount of abstraction was undoubtedly involved in the result.

As one compares a column of Arabic numerals with an ancient Indian set of numerals, the differences seem extremely slight — thus confirming the Hindu priority for the invention.* Al-Khwarizmi's achievement amounted to his recognition of the elementary visual simplicity of the Hindu system plus the demonstration of its practical usefulness. The manipulations of the Indian mathematicians had often been shrouded in mystical abstractions. Their texts are often obscure, presented in verse or purely rhetorical form, sometimes in lovely poetic language. (The one outstanding exception, Bhaskara's clear and systematic *Lilavati*, was written some three hundred years later than al-Khwarizmi's work, and was undoubtedly influenced by the Arabs.)

In his *Algebra*, al-Khwarizmi (whose name derived from the Persian province of Khorasan, where he was born late in the eighth century) set forth clearly intelligible demonstrations of the uses of the positional system, with examples of equations, multiplications, and divisions — including some discussion of the principles of squares and roots — the basic operations on which, in one way or another, any more complex computations are based. As far as it was an interpreter's work of the Hindu *Siddhanta*, the interpreter's achievement consisted of his grasp of the essential and his penetrating recognition of its multiple potentialities.

One of the phenomena that particularly fascinated al-Khwarizmi was the equation, with its dynamic possibilities. By using the word *al-jebr* (the reducing and recombining of parts), he had acknowledged the fact that in an equation one adds or subtracts quantities of identical magnitude on both sides — treating it like a scale, as it were, that one wishes to keep in a perfect balance. He clearly saw the potential of the equation as an incomparably sensitive calculating tool, with whose help, for instance, the builders of Gothic cathedrals were later able to plan their bold experiments in the distribution of large structural weights.

* It is interesting that when Leonardo of Pisa introduced the Arabic numerals to the West, he explicitly called them "the nine fingers of the Indians," giving credit where credit was due.

It would probably be too much to claim that he actually sensed the role equations were to play in some distant future in solving advanced mathematical problems — such as problems in the relationship of inertia and acceleration as a basis for determining motion; or Newton's use of the equation in the differential and integral calculus; or the equation's use for defining the motion of air and fluids by the eighteenth-century mathematician d'Alembert. What is certain is that al-Khwarizmi recognized the inherent philosophical principle — the capacity of the equation to define complex relationships by establishing an equilibrium between quantities of a virtually unlimited scope, its ability to define unknown factors ("x") through the principle of the "scale."

Al-Khwarizmi did more than grasp the principle; he gave Western thought the first examples of equations, neatly classified according to basic problems, as a springboard for any future, more sophisticated, use. That among these he even included a geometrical demonstration along Euclidian lines may serve to highlight that historic fusion of the Greek and Hindu mathematical traditions epitomized by Islam.

* * *

None of these lessons was learned overnight. It was not as though Medieval Europe actually "went to school" with its Islamic teachers, or that Europeans developed their science simply reading the right books (in their defective translations), doing their homework like good little kids. Even if historians, with their professional exaggeration of the importance of the written word, tend to view books as the sole agents in the diffusion of ideas, this is obviously not how one culture influences another.

In reality, such influences from an advanced to a more backward civilization come in a vast, diffuse torrent, with the waters carrying along a good deal of mud and debris. If the translations contained serious distortions and misunderstandings, more misunderstandings arose at the other end. A comparatively obscure Islamic philosopher in twelfth-century Spain, ibn-Rushd (whose name was westernized into Averroës by the translators), caused an enormous upheaval among the intelligentsia of Europe. His philosophy, centering on the idea of the universal "unity of the mind," became a fighting platform for anticlerical opposition, a program for free critical thought unhampered by theological doctrine. During the 1270s, a century after he actually wrote, radical students and faculty members at Paris were defying traditional teachings under his name. Dante, who belonged to that younger generation (though as a student in Florence), later placed Averroës among the great philosophers of antiquity — in the Inferno to be sure, but only in Limbo, along with other pagans. Yet Dante singled him out with the poetic tribute *"che il gran commento feo"* (he who wrote the great commentary); namely, on Aristotle's philosophy.

Alarmed, Church and academic authorities scurried around for measures to stem this dangerous trend. Averroës' teachings were condemned in an explicit list of two hundred and nineteen "errors." The Dominican order dispatched Thomas Aquinas to Paris in the hope that he would smoothe the anticlerical wave with his conciliatory philosophy. Yet, as the positions polarized further, even Aquinas' reasonable stand began to seem far too liberal to the frightened conservatives (and too sympathetic to Aristotle, on whom Averroës had based his provocative thought). The "Angelic Doctor" himself was about to fall victim to the eternal peril of those who preach reason when two camps are confronting each other in growing hostility — the danger of being torn to pieces by both sides. Aquinas left Paris with the shadow of theological condemnation hanging over his name, as it continued to do for years after his death. In 1277, his aged teacher, Albertus Magnus, had to rush to Paris from his German retreat to defend both Aquinas' memory and their common beliefs. But Aquinas remained *persona non grata* to the Church authorities way into the fourteenth century, and some of his teachings were, in fact, expressly condemned.

At least he was spared the fate of some of the rebel leaders, who were burned at the stake, or of the chief spokesman of the Latin Averroists (Latin as distinguished from Muslim), a brilliant young philosopher by the name of Siger of Brabant, who was hounded from Paris and later assassinated under mysterious circumstances somewhere in the south of France.

Averroism continued to spread through the following two or

three centuries as a somewhat nebulous philosophical movement, but always with an electrifying aura of radical thought, connoting a spirit of uncompromising rationalism, in particular in matters of science.

* * *

Misunderstandings, misplaced emphases, overreactions. Cultures ultimately respond to each other very much like individual human beings. Islamic science was not simply accepted for its own sober sake, word for word, book for defectively translated book, subject for intricate subject. Instead, it seemed to carry heavy ideological overtones, arousing deep passions for as much as against its implications and substance.

Aristotle's scientific writings were banned at the University of Paris as early as 1210 by an express decree of a Provincial Synod, reiterated by various ecclesiastical bodies at different times during the thirteenth century — which was probably the reason why his Western admirers, for the time being, chose to concentrate on Aristotle the philosopher (which is how he was still treated by Thomas Aquinas, despite Aquinas' undoubtedly thorough familarity with Aristotle the scientist).

It took over a generation before a few courageous teachers, like Roger Bacon, began to sneak bits of Aristotelian science into their classroom curricula — a provocative step that, after all this long-standing suppression, may have contributed to the eventual ideological militancy of the Averroist storm. Yet, even though Aristotelian science was initially forbidden at Paris, the recently founded University of Toulouse merrily advertised, in what we might now call its "catalogue" for the year 1229, the "teaching of the books on natural science that have been banned at Paris." Aristotle, Averroës, Greek and Islamic science had become magic terms, causing nightmares for one group of people, excitement for others, depending on how one felt toward modern ideas and their unsettling impact on fixed intellectual ways.

As a matter of fact, even the Arabic numerals and the new mathematics they engendered did not hold their triumphant entry on the European scene merely because someone translated and others faithfully studied the proper books. The victory of sheer, beautiful rationality over far less rational methods was delayed by the hopeless irrationality of men. Arabic mathematics penetrated only intermittently and with astonishing slowness, taking several centuries to sink in, although in this instance one interpreter's methodical explanations happened to play a decisive part.

European merchants who had for centuries been in lively business contact with the Muslim world must have had a pretty good idea as to how their best customers handled their accounts; it would be very strange indeed if some smart businessman had not adopted the Arabic system. Nor had Arabic arithmetic escaped the sharp eye of the major twelfth-century translators: Gerard of Cremona wrote a short treatise on algorisms (a manuscript copy is still at Oxford's Bodleian Library). But at least this particular effort by the great translator seems to have remained without real effect. An actual methodical explanation, and therewith the effective introduction of Arabic mathematics to European scientists, came comparatively late, with Leonardo of Pisa's *Liber abaci,* which was a careful exposition of al-Khwarizmi's work and was initially published in 1202 under a slightly Latinized form of its Arab title.*

Leonardo, who had grown up in North Africa, where he got to know the Arabic system (his father was stationed as a customs officer there), faithfully proceeded along al-Khwarizmi's lines, giving examples of equations, stressing the usefulness of geometric demonstrations (and thereby introducing a first seed of the idea of the interchangeability between geometric and arithmetic expressions in Europe, a result of the Greek-Hindu fusion that al-Khwarizmi had accomplished), and generally setting forth the principles and possibilities of the Arabic numbers.

Slowly, the Arabic system penetrated into commercial and mathematical use. Mathematicians like Jordanus Nemorarius ingeniously operated with it during the early thirteenth century. The astronomical tables of Alfonso X of Castile, known as "the Wise," expressed Islamic observational data in Arabic numbers;

* Leonardo called his translation *Algebra et almuchabala.* Al-Khwarizmi's original title had been *al-jebr we'l mukabala,* meaning the "restoration" (*al-jebr*), that is, of the equilibrium, by adding or subtracting the same amounts on both sides of an equation, and the "simplification" (*al mukabala*), that is, the combining of equivalent terms into a single expression. But Europeans understandably preferred the simplified *Liber abaci.*

and a scientist like Roger Bacon, with his consuming interest in questions of methodology, wrote about the Arabic notations and emphatically recommended their use. All that, however, was an extremely gradual affair. A full replacement of the Roman numerals did not happen until the Renaissance, and Renaissance artists still preferred to date their paintings often enough by using the decorative Roman symbols, which no doubt appealed to their preference for classical antiquity.

Neither did science arrive in a single, neatly labeled package. It came as part and symbol of a more sophisticated culture, and this general cultural context rubbed off on the meaning of science itself. The very word *"scientia"* carried a different connotation from the one it carries for us (though we may have inherited some of the ideological overtones). It meant urbane sophistication, up-to-date education — in a word, "knowledge." For centuries prior to our own hyperspecialized age science possessed some of the earmarks of an intellectual fad. Humanists in the fifteenth and sixteenth centuries still liked to dabble in scientific (or pseudo-scientific) problems — whether in astronomy, astrology, geography, mineralogy, alchemy, zoology, botany, or what-not — happily mixing them with their love for the ancient philosophers, historians, and poets. To a Europe steeped in the adventure of rediscovering the earth, the study of nature was an intellectual delight, not necessarily a stern and specialized academic pursuit.

Besides, science arrived in the cheerful company of Arab love poems and prose, celebrating an erotic concept of the relationship between the sexes, provoking upheavals in the grim feudal attitude toward women and sex. European chivalry and creative writing were as much influenced by Islam as was European science. The decorative arts, manuscript illuminations, tapestries, furniture, architectural features — all these finer things of life for the enjoyment of the privileged classes were reflecting the Muslim touch.

Europe received Arabic science by no means for its own sober sake alone, but very largely as part of a cultural movement for which Europe was ready and to which all its energies were attuned. There is a glitter, a glamour, a glowing and colorful radiance emanating from the Muslim countries across Europe in the final Medieval centuries that has more to do with a liberation of the senses than the mere diligent study of optics or arithmetic. Yet for a few still rather isolated and often lonely minds, science happened to mean precisely what it had signified to the scientists of Islam: a type of highly specialized inquiry that they were ready to take up where the Muslims had ended and carry forward on its exact and exacting terms.

FIVE

Scholastics, Mystics, Alchemists

Natura parendo vincitur
(Only by doing nature's will
can we hope to control it).
Roger Bacon

SOMEWHERE there is a discrepancy in our picture. It all seems
consistent enough. During the early centuries following the
collapse of Rome, the mind had turned from the observation of
nature and immersed itself in those vast metaphysical dimensions
of the early Medieval view of the world. Then came the slow,
steadily growing pull of this world as technical know-how began
to produce significant agricultural progress, as the surplus crops of
the land generated a thriving trade, trade brought forth early capitalism, and the early capitalist towns began to flourish in their
youthful vitality amidst an as yet largely Medieval scene.

Logically enough, these profound social changes had opened
the way for intellectual adjustments and signaled the gradual return of the mind to earth: the School of Chartres proclaiming the
study of nature; Chartres's new natural cosmos filling up with the
details inherited from Islam; the historic linkage with the scientific legacy of the ancient world. At last, the breakthrough of
an autonomous Western science, operating with the materials
amassed by preceding cultures, yet using a dynamic thrust all its
own — a probing intellectual agility, a critical acumen, a methodological self-assurance — all fruits of the vigorous discipline of early
Medieval thought.

Newborn Western science did more than organize, criticize,

probe: by carefully logical stages it refined, revised, and in the end overthrew humanity's age-old ideas of the cosmos, replacing them with the modern solar universe with which science still operates today. As the hazy outlines of the cosmos of Chartres emerged into more detailed focus, as the universe stood out more and more sharply, a solid, mathematically calculable firmament spanning the earth, science's glances increasingly turned to the earth itself. Explored to its distant corners by the age of discoveries, remolded into a near-perfect sphere, the new earth at last became the conceptual "building block" with which the Scientific Revolution was able to construct its sun-centered universe, with its literally infinite possibilities and implications. Historically, it was the new concept of the earth that served as a spearhead into the modern cosmos, in which earthlike planets were whirling around.*

Naturally, these momentous theoretical developments were accompanied by other manifestations of a great change in the outlook of the later Middle Ages and Renaissance: a profound philosophical reorientation involving a radical revision of the basic direction of thought; a growing earthiness in poetry, literature, and art; and a definite ascendancy of the visual arts over other forms of expression — marking a phase in which the sheer visual experience was to take on an almost ideological significance, defying those purely spiritual orbits in which the Medieval mind used to dwell. The discovery of the earth as a phase in theoretical science was accompanied by a growing earthiness in perception and thought.

* Though we usually do not think of Medieval science as leading into modern with any significant degree of continuity, the connecting tissues are, in fact, so vital that they suggest a continuous, organic process. This is true for the methodological premises that were, in the main, worked out by the later Middle Ages, including the new conceptual approaches to the problems of motion and space. (See A. C. Crombie's concise summary in his chapter "The Continuity of Medieval and 17th Century Science," *Medieval and Early Modern Science*, New York, 1959, volume II, especially pages 106 and following.) Continuity is equally significant in cosmological thought, where the steady expansion of astronomical data by Islamic and Medieval European observers led to the eventual discarding of the geocentric and its replacement by the heliocentric theory of the solar system. At the same time, the new concept of the earth that emerged from the age of discoveries shattered one of the major foundations of Aristotelian physics and thereby ushered in the conceptual revolution that was climaxed in Newton's universe. (See my "Renaissance Concept of the Earth in Its Influence upon Copernicus," *Terrae Incognitae* [*The Annals of the Society for the History of Discoveries*], volume IV, 1972, pages 19–51.)

Closer to the domain of science there had been parallel developments in technology, more palpably prompted by early capitalistic needs: the search for new sources of power, the creation of a glittering array of mechanical constructs — all serving to augment human productivity, ultimately to enhance human mastery over the hidden forces of the earth. The rise of science during the later Middle Ages was without doubt an integral aspect of the change from a feudal to an early capitalistic society, and of the general expansion of consciousness which that profound social change brought about.

* * *

The flaw, the discrepancy in this apparently logical and consistent picture, is that it seems just a trifle too logical. The great charm of the Middle Ages, we would think, lies in their mysticism and magic — in the intimacy that is generated by the narrow crooked streets brooding in the early morning sunlight or resounding to the noises of labor; of streets steeped in the smells of food and wine from the taverns; of a life that seems at once earthy and mysterious. The tolling of church bells, the cool, dim interiors of the churches themselves; the feeling of reverence before another world as one enters a Medieval cathedral — changeless and filled with some higher knowledge, of which we seem to have forgotten the name — these things do not readily suggest the sober workings of the scientific mind nor the infallible logic of economic causality. Somehow, our picture of the Middle Ages does not have room for all that much consistency or rational thought.

Nevertheless, the evidence seems to speak with an unequivocal voice. Historians may argue over this or that particular interpretation, but not only does the whole process strike us as intrinsically logical; the type of scientific thought that developed under these stimulating conditions is marked by its distinct inner coherence and powerful rationality. The vigor of rational inquiry and the speculative force of theoretical thought are unmistakably present in the scientific manuscripts of the later Middle Ages and Renaissance, a continuous source of amazement for the modern historian. And the inner coherence of the thought process leaps to the eye as one tries to "read" these documents in their logical sequence, reconstructing the conceptual threads by which they are linked, and relating them to the larger historical context. A straight line of rational thinking leads from the assimilation of the Greco-Islamic heritage, through the physical and astronomical speculations of fourteenth-century theorists, to the bold insights of the Scientific Revolution. Once more, the modern reader is taught an impressive lesson in humility: the Medieval scientists pondered their problems with unexpected intellectual force and handled them, step by step, with perfect consistency.

Despite all this rationalism, one could not doubt the essentially mystical quality of Medieval life. Almost every remnant suggests a climate that was far removed from our sober-minded modern practicality — a culture absorbed in grave spiritual matters; the stark contrasts of a vision that saw the events of this earth not for their isolated and limited ends but as symbols in a great metaphysical drama, taking place somewhere outside the narrow confines of human design. The problem is how such heterogeneous elements could have possibly lived side by side — the sturdy pragmatism of the economic and technical advances (with their evident reflection in culture and thought), the rationalism and inner consistency of theoretical science, and the dreamy atmosphere of a civilization based on a mystical conception of the world.

* * *

We are facing a problem in cultural history. Patently, cultures are not simply controlled by a single dominant feature. There has never been a culture whose life evolved in a single, simple "key," no matter how much it might look that way to our foreshortened retrospective vision (which is always apt to blur the nuances and to simplify the subtler shadings of actual life).

Nor, for that matter, is science always the product of the same, invariably consistent, type of rational thought, regardless of the cultural context. Different cultures at different times have made their contributions to our knowledge of nature on methodological and philosophical grounds that often were far from rational and may seem utterly unacceptable (sometimes even ludicrous) to the modern scientists. Still, what counts for the progress of science is solely the quality and nature of the contribution, not the shifting cultural context from which that contribution may have been made.

The truth is that the Middle Ages made their contribution to science from a rich combination of intellectual traditions and cultural attitudes. Mysticism and magic presented as fertile a soil as did pure rational thought. If, offhand, we may take it for granted that science must have been the exclusive product of a rationalistic approach, we would in effect merely project our modern experience — in which science and rationalism are indeed linked, like Siamese twins — back into an earlier and decidedly different cultural context.

The true complexion of the intellectual life in which Medieval science developed differed so radically from everything we now associate with a scientific climate that it defies our comprehension, unless we concede that each culture functions in its own distinct way and produces its science strictly under its own peculiar philosophical assumptions. To grant this premise would seem highly worthwhile; without it, we would shut out all those metaphysical, mystical, and magical elements that lent Medieval science its characteristic spice and its ultimately irrational (and, just for that reason, intensely colorful) atmosphere.

* * *

A capacity for keen rational thought was by no means alien to the Medieval mind. It was due to just that long-ingrained habit that Medieval science was able to surpass the scientific achievements of Islam — and even from the very beginning to approach the body of Islamic data with a superior sense of methodological order and intellectual discipline. The Medieval mind had acquired its remarkable skill in logical argument, in the uses of precise rational categories, and the handling of abstract concepts ever since the time of the theological disputes of the early Church. The subtleties of the early dogmatic disputes — the Arian, the Filioquist, the Donatist controversies, and whatever else their manifold names — were the original training ground for Medieval rational logic.*

* The issues behind these early Church controversies were the nature of Christ (or His relationship with God) and the role and authority of the Church. Both issues seem fundamental enough to explain why they aroused such vehement passions during the early centuries (the Donatists caused riots and street fighting

The Medieval mind had cultivated these talents across the whole spectrum of transcendental speculation in which Medieval philosophy had so intensely — and fruitfully — indulged. But if an emphatic rationalism was therefore entirely endemic to traditional Medieval thought (and was further refined, almost to a modern level, during the later Middle Ages, mostly under the influence of Aristotle's translated writings), at no time was it the only, nor even necessarily the dominant, intellectual trend. Nor was science invariably wedded to the rationalist tradition.

It would have been perfectly common during the high Middle Ages to find eminent scientists, capable of the most logical theoretical thought, nevertheless living within their own weirdly mystical world. Among their number we would have encountered Roger Bacon, the man who probably possessed the keenest grasp of the nature of science in the Middle Ages. Some might staunchly proclaim their commitment to a rationalist philosophy, yet when it came to such intriguing practices as alchemy or astrology, we might find them merrily dabbling in the old magic. So strong was the hold of these time-honored "arts," as they were called, that they were still being practiced — and indignantly criticized — among the eminently enlightened humanists of the Renaissance.

Nor, for that matter, did rationalism necessarily act as the unfailing source of exact, straightforward, trenchant thought that we would expect. Under the impact of Scholastic teachers, who had gained virtual control over the academic establishment by the second half of the thirteenth century, rationalism sometimes took on a pedantic, rigid, excessively abstract face, which, often as not, could stifle fresh scientific ideas. Far more original initiatives might, in the meanwhile, originate in the mystic camp. Renaissance humanists from Petrarch to Erasmus were to react with fierce anger and ridicule against the unworldliness of the

in the North African cities during the fourth century). If early Christianity had given in to the Arians, who denied the divine nature of Christ, it would have lapsed back to a monotheism close to the Jewish tradition; if the Church had yielded to the Donatists' demand for uncompromising purity of the faith despite the pressure of persecutions, it would have surrendered its claim to universality for the exclusiveness of a sect. Both "heresies" (together with others) were combated with trenchant theological arguments, the Arian primarily by Athanasius at the Council of Nicea, the Donatist by St. Augustine.

"Schools" (meaning the academic establishment under Scholastic control), denouncing a mechanical type of rationalism that could turn even the study of nature into a pallid intellectual game.

A rationalism capable of frustrating genuine intellectual progress; a mystic tradition that not only could interfere with clearheaded scientific thought (as it often enough did), but might actually work as a major source of scientific insights; a culture bewilderingly mixed of mystic and rationalist strains — such are the puzzles of intellectual life at the height of the Middle Ages. What had brought about this paradoxical situation?

* * *

The civilization of the high Middle Ages, with its stark contrasts and contradictory elements, was the product of two substantially different evolutions. Medieval science was clearly indebted to both.

That brisk climate of progress that nourished the technological experiments, promoted an early industry, gave rise to an energetic building activity in the towns, and more and more left its imprint on writing, art, and thought was the natural outgrowth of an early capitalistic civilization. As for capitalism itself, it was the logical product of the dynamic frontier society of the early Middle Ages spurred to extraordinary efforts by powerful odds. On this down-to-earth level of economic history and its immediate effects, matters are refreshingly clear. What we face here is little more than a straight process of impressively successful recovery against staggering odds, from centuries of chaos and desperate misery to the prosperous life of the cities. The most refreshing element for one who studies the Middle Ages is that the period is built around a spectacular "success story," from frontier to urban prosperity.

The complexities of the situation arise from the fact that these elementary economic developments had taken place within an amazingly intricate structure of social organization and cultural heritage. Both were the result of much earlier experiences, going back to a time long before economic recovery had overtaken the scene — the time when the mind had sought refuge from the collapse of Rome in the metaphysical realms of St. Augustine's "City of God"; and when a coarse military caste had saved the West, after much turmoil and suffering, from drowning in chaos by establishing the amazing phenomenon of the feudal society. Both social organization and cultural mentality were, therefore, gigantic historic anachronisms, which does not mean that they were any less resilient or powerful. The great tensions of Medieval civilization at its height derive from the relentless conflict between these two forces — the frantic adjustment to the breakdown of ancient Rome in social organization and cultural attitude (which really meant the perpetuation of that traumatic shock), and the dynamic vitality of economic resurgence. Modern science not only was born from that conflict but was indelibly stamped with its inherent tensions and strains.

The culture of that peak period — an age of brilliant creative feats — speaks to us in eloquent language of the inner struggles with which the individual had to contend: the nervous thrust of the Gothic cathedrals; the tortured love songs of troubadours; the tense bodies and emaciated faces of statues adorning the cathedral façades; the agony of Dante's tragic young love, and the subsequent search of his tortured soul, with his own personal hell, catharsis, and final salvation projected onto a cosmic backdrop.

The human reality behind these creative achievements with which high Medieval culture abounds was one of conflict. It was a conflict precipitated by the contradictory forces of history and experienced in the suprapersonal terms of an entire culture. Nevertheless, each individual had to suffer it through in his or her own mind and soul. Women were very much included, as we know from, among other evidence, the heart-rending letters of Héloïse to her unfeeling lover, the famous Abelard, himself a pioneer of philosophical thought.

The individual conflict was between the time-honored Medieval tradition of world-denial — the psychic legacy of the collapse of Rome — and the new social realities that were opening tempting glimpses of the here and now, the contemporary, more and more prosperous world, which was, in turn, opening up a vast prospect of the world of nature. Behind the tense sinews and strained facial muscles displayed on the statues of saints lies a culture's fierce agony between present and past, between human nature's elementary needs and the taboos of an austerely restrictive tradition. If the present beckoned to one to relate to the world through the full use of the senses, one's upbringing dampened any such exuberance by reminding one of a tragic past, with its lesson of turning from the world in anguish.

Caught in that impossible dilemma, which was continuously sharpened as the growing prosperity increased the lure of the world, the Medieval mind proved its remarkable resourcefulness by devising a variety of escapes. One was the escape of sheer artistic creativity, which has given us the precious cultural legacy of the later Middle Ages. Another was loftily intellectual — the great philosophers' formidable attempts to wrestle rationally with the implicit problems. (Thomas Aquinas' great universal system contains a series of profound answers to the metaphysical questions raised by the dilemma of the times.)

Mysticism represented perhaps the most typically Medieval response — the attempt to solve the dilemma by escaping into that supranatural sphere in which the Medieval mind felt most completely at home and where all the conflicting forces could coexist harmoniously. Medieval mysticism, greatly in vogue during the thirteenth century, became the common denominator — more truly, the common dimension of feeling and thought — in which all those harrowing tensions and contrasts of the real world were all of a sudden blissfully reconciled: the traditions of the past and the temptations of the present; philosophical thought and creative art; world-denying spirituality and the lure of nature; even science and faith.

Medieval mysticism meant accepting the rule of invisible forces, rooted in the beyond, over the tangible, everyday experience. Mysticism countered downright practical thinking with poetic feeling, the pragmatic acceptance of things as they are (or as they present themselves to our senses) with a profound awe before their role within the Good Lord's mysterious blueprint.

For the mystic, the wind moving through the foliage of a tree is not just a force of nature acting on a member of the vegetable kingdom, but God's finger playing lightly across the chords of a marvelous harp. A flock of birds shooting across the evening sky may stir the mystic to ask why it is that the movements of humble creatures are endowed with a more haunting beauty than the most exquisite work of art. Observing some leaves of grass, the mystic may wonder at the life force that pushes them through the crust of earth every spring, and neglect more specific questions.

It was not so much rationalism that contested the mystic view of the world — after all, impressively logical arguments, depending on one's premises, could be made in favor of either point of view — but an earthy, pragmatic attitude, an unflinching positivism and empiricism, that assumed without a great deal of philosophical inquiry that what we observe with our senses must take precedence over the invisible powers of the beyond. (Only where rational thought was wedded to this pragmatic approach do we have "rationalism" in the modern sense of the term — the kind of rationalism that we have now come to associate with science.)

* * *

Mysticism, irrational and intrinsically "unscientific" though it may seem to us, was able to produce valuable scientific insights. This may be in part because science is ultimately a form of creative activity, and the mystic approach may be extremely stimulating for the creative powers. After all, in all creative activity motivations are in the final analysis irrelevant. Moreover, Medieval mysticism, by the very fact of its poetic approach, involved obvious incentives for the study of nature — even though the mystic's inherent sense of wonderment and poetic love had to ally itself with the virtues of patient observation to net valid scientific results.

There is still a more remarkable factor. Sufficient uncertainty surrounds the ultimate philosophical premises to admit of at least a reasonable possibility that those invisible (and often unverifiable) forces may in fact exist and exert their mysterious influences on our everyday environment. Patently, modern science has come around to accept some curious phenomena that the more literal-minded nineteenth-century scientists would have laughed off as humbug but Medieval mystics might have accommodated without any difficulty at all: the manifestations of extrasensory perception; the varied signals and intrusions of the unconscious of modern psychoanalysis (perhaps even the "collective unconscious" of Carl Gustav Jung); or, on a somewhat different plane, modern scientists' growing suspicion that our customary premises and modes of thought may break down before certain, as yet unexplored, physical phenomena — whether subatomic or supragalactic— conceivably signaling basic deficiencies in our methodological approach. Medieval mystics took it for granted that unknown forces are acting on us from somewhere outside (or indeed inside) ourselves; they would scarcely have been surprised to find our meticulous pragmatic rationalism faltering before a deeper comprehension of the world.

Perhaps these, and a thousand other symptoms, may suggest that the pragmatic rationalism (or "rationalist positivism") of a more confident phase of our modern age has run its course, that Medieval mysticism may well have contained some grains of a higher wisdom — besides a rich dimension of the experience — that has been essentially closed to us ever since. In science, at any rate, the Medieval mind worked in a way that was able to combine the mystic with the pragmatic approach, untested magic beliefs with straight empirical observation.

* * *

Rather than think of this peculiar combination merely as a transitional phenomenon in which promising new beginnings were still deplorably mixed in with obsolete "superstitions," we should try to understand the unique cultural atmosphere that enabled both elements to produce valid scientific results. It is mysticism's creative ability (rather than people's slowness in shaking off obsolete ways) that explains why the mystic approach retained its influence straight into the Scientific Revolution. Such modern pioneers as Kepler or even Newton are apt to startle the modern reader by their sudden flights into mystical spheres; they are, in effect, paying a tribute to the intellectual stimulation of their Medieval predecessors.

This odd mixure was present at every turn. During the thirteenth century we find men like Albertus Magnus haughtily rejecting occultism and magic. An important pioneer of empirical observation himself, an eager observer of plant and animal life as he traveled through Germany and parts of Northern Europe in his capacity as Provincial of the Dominican order, he seems the perfect embodiment of clear-headed modernity in an age still involved in all manner of magic beliefs. But lo and behold! The great Albert himself could at times dispense the fanciest notions — that a lion's tooth when hanging from the neck of a boy before he loses his first teeth will protect him from toothache by the time his second teeth push through; that lion's fat mixed with other oint-ments will remove blotches; that a lion's brain, when mixed with a strong oil and inserted in the ear, is a sure cure for deafness. (One wonders whether lions may not have been in short supply for all these therapeutic applications!)

Perhaps one might expect such inconsistencies in the thirteenth century; but some three hundred and fifty years later, in the very midst of the Scientific Revolution, with the Age of Reason already well under way, Johannes Kepler still presented his world-shaking new theories about the motions of the planets in a mystical lan-guage and cast of thought that make his writings sound like typ-ical Medieval treatises. By the end of the seventeenth century, Sir Isaac Newton formulated his revolutionary conception of the physical universe, based on the most advanced mathematical thinking, in his *Philosophiae Naturalis Principia Mathematica*. Yet for twelve weeks, even as he was writing this pioneering work, he kept several alchemical furnaces going in search of "something more noble, not to be communicated without immense danger to the world, if there should be any verity in the Hermetic writers," as he was to state in another context. Indeed, the great Isaac New-ton, father of the modern scientific universe, devoted a major part of his life to the pursuit of alchemy (for which the Hermetic writ-ings were considered basic). Large portions of his copious unpub-lished manuscripts deal with this mystical subject.

A deplorable Medieval hangover, to be passed off with a smile? The case is not quite that simple: Newton was in frequent contact with Robert Boyle, one of the founders of modern chemical sci-ence, and seems to have shared with Boyle some of the crucial thought that led to the transforming of the mystical alchemy of the Middle Ages into the beginnings of modern empirical chemis-try. It was not just that mysticism and magic took a long time to die; before they did, they bequeathed their full positive legacy to modern science, as an immensely vital heritage and a signifi-cant source of inspiration. In fact, the continuity is so direct that historians trace the origins of modern scientific chemistry, in many vital aspects, to the Medieval alchemists, who themselves belonged to a tradition reaching back to the earliest civilizations. Newton dabbled in alchemy because his genius urged him to share in the transformation of that Medieval tradition into modern chemistry — even as he took the lead in the emergence of modern physics, modern mathematics, and the modern scientific universe.

* * *

Medieval scientists made a solid contribution to the development of astronomy. They did so by critically assimilating the cosmological theories they took over from the ancient world (largely through the medium of Islam) and by expanding the observational data on the course of the stars, which were recorded for the benefit of sailors as well as scientists in the Alfonsine Tables, an important collective enterprise initiated on the basis of Islamic sources by King Alfonso the Wise.* The new sun-centered theories of the Scientific Revolution would have been unthinkable without this body of empirical observations and its gradual expansion and refinement by a number of astronomers between the thirteenth and early sixteenth centuries.

Yet, at the same time, the stars were believed to rule the course of human life, the occurrence of illnesses, the fate and nature of animals, minerals, plants. For thousands of years astrology and astronomy had lived peacefully side by side. By the thirteenth century the new brand of rationalists — such as Albertus Magnus — were turning vehemently against the presumptuous magic of the astrologers, using a combination of rational and religious arguments. Paradoxically, even scientific astronomers continued to use the data compiled by the astrological magicians — who, with pointed hats, their long gowns sprinkled with the symbols of stars, minutely calculated the stellar movements to determine a horoscope, a beneficial stone, a favorable flower, a healing herb. Even Albertus Magnus (who rarely matched his rationalist fervor with particular methodological consistency) advocated the use of astrological data for the purposes of astronomy. And why not? To discard this rich lore would have meant throwing the observations of the ages to the winds, merely because their premises failed to com-

* The Alfonsine Tables were completed around 1270, but only gradually replaced the so-called Toledo Tables, computed by the astronomer al-Zarquali for the meridian of Toledo, or a series of Latin adaptations of that Arabic work, based on the meridians of Marseille, Paris, Pisa, Palermo, and London respectively (J. L. E. Dreyer, "Medieval Astronomy," reprinted in *Toward Modern Science*, edited by Robert M. Palter, New York, 1961, volume I, pages 243 and following, and page 252).

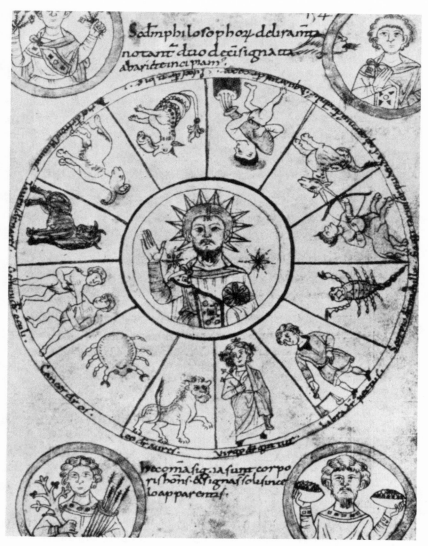

Zodiacal chart from the tenth century, with the figure of Christ in the center surrounded by the signs of the Zodiac, each for a particular part of the body, epitomizes the astrological tradition against which thirteenth century rationalists were turning with occasional vehemence.

ply with the more recent, more rational scientific objectives. If Albertus was lacking in consistency, on a purely pragmatic level he was obviously correct.

We are witnessing the beginnings of diversification within the natural sciences. The process included the branching out of the modern empirical discipline from the traditional mystical root: astronomy striking out on its own from the astrological mother science; chemistry from the traditional body of alchemy; at some point — roughly midway between the first stirrings of modern astronomy and the emergence of chemistry — geography parting company with cosmology, with which it had coexisted (under somewhat confused conceptual notions stemming from the Aristotelian system of physics) until the fifteenth-century Renaissance.

True, in this newly found assertion of their independent identities, the early modern sciences could often follow ancient models: Hellenistic science, beginning with Aristotle, had shown a definite trend toward diversification; and the writings of Ptolemy toward the end of antiquity involved a clear concept of astronomy, as distinct not only from its astrological forebears but from geographic science as well (on which Ptolemy wrote his *Geography*). But as the Medieval mind, under the guidance of Islam, retraced the ancient road of gradual diversification (anticipating the Renaissance in this respect, too), it set the definitive patterns for the modern age, creating the specialized sciences that were henceforth to stand on their own. It was with these new departures from the traditional (and quite often characteristically mystical) root sciences, in the direction of clearly defined empirical disciplines, that the shape of modern science received its mold.

As part of the inner logic of this process the revolt against mysticism was to run its course on methodological, rather than on substantive grounds. Despite all those angry denunciations of the magic "errors" and their practitioners, the mainstream of science preserved a good deal of the mystic substance into the early modern age. Where the antimystic rebellion focused its fiercest and most effective attacks was on questions of method and approach. Eventually, mysticism was driven underground, so to speak, and the modern scientific approach came to vaunt its clear-headed rationalism and empiricism. In a very real sense, the modern concept of science (and of what we consider "unscientific," or, in other words, false) was developed when the methods and approaches of the alchemists and astrologers, notwithstanding their rich substantive contributions, were overcome. The revolt against otherworldliness was to leave its strongest impact on the general attitude of the Western world.

* * *

There is a good deal of mysticism in the lore and the primitive magic of Medieval medicine. It is easy to say (as most textbooks do) that Europeans "learned" medicine from the Arab world, mostly by studying the ancient writings of Hippocrates and Galen and their Islamic commentators and, around the time of the Renaissance, began to practice dissection and serious physiological studies, whereby modern medical science was successfully launched. Such generous outlines are easily sketched, leaving the impression that the whole process was accomplished with the dreamy ease of a fairy tale. They also imply, once again, that it was all a totally rational affair, little more than a question of reading the right kinds of books.

But people continued to get sick, sick people were healed, the medicinal properties of herbs and plants were explored, the infirmaries of monasteries were serving as hospitals (before the institution of urban hospitals was taken over from Islam). Monks and nuns were acting as physicians, barbers as surgeons, a body of lay physicians gradually came into being, pharmacists — under Islamic influence, to be sure — began to open apothecary shops in the streets of Medieval cities and to organize themselves in guilds.*

Hippocrates and Galen? Of course: their works were taught at the oldest Medieval university of which we know, at Salerno, and a bit later (by the twelfth century) at the newly founded University of Montpellier in the south of France. But they were mostly taught textbook fashion, in the form of straight abstract theory, with little actual dissection or demonstration. It was a long and

* The Medici family, for example, took its name from the Florentine guild of physicians (medici, in Italian) and pharmacists, which they joined during the thirteenth century; and its coat of arms, the three gilded balls, from the pills that were the pharmacists' trademark. (The story that the modern pawnshops have perpetuated these origins in their sign, reflecting the banking activities of the Medici, has been seriously questioned.)

circuitous road before such purely theoretical knowledge could be put to work to make a sick person well, for the Greek texts (and their Arab commentators) were shot through with contradictions. Their theories, what is more, often rested on philosophical premises that were unknown to the Medieval West. Before they could be applied to the treatment of patients, Medieval physicians needed dictionaries that would explain alien concepts or terms, commentaries elucidating a difficult text or reconciling some divergent opinions. And this sort of explanatory literature did come forth by the ream, bridging — and at the same time underscoring — the gap between Greek medical theory and its practical application in the day-to-day fight against Medieval diseases.

How then did Medieval doctors practice their medicine? Largely through a mixture of elementary empiricism and mystical wisdom, something like a sixth sense for the nature of illness — and, even more important, for the healthy forces inherent in human nature, *"vis medicatrix naturae,"* nature's intrinsic healing powers (to use the Hippocratic term). In short, the physicians were using precisely the same instinctive sense that makes for a good doctor now as compared with a mediocre one, who has merely completed the prescribed training. Or, we might say, they were relying mainly on a "spiritual" approach to illness, the kind of approach that may now be practiced by the adherents of Christian Science, or that underlies the more explicit assumptions of psychosomatic medicine. The crucial difference between the Medieval and modern practice was that such instinctive feeling was deliberately fostered in Medieval culture (rather than being treated with embarrassed silence because of its "unscientific" quality, as in ours), and that it was all the more effective for being consciously accepted and applied.*

In time, the fame of the School of Salerno spread from its site to the south of the Gulf of Naples, mostly because of a number of treatises on medical, dietary, and hygienic lore attributed to the

* By reviving the great Hippocratic tradition, Medieval medicine was, in other words, committed to a humanist approach, one that started from the total human person, including his or her natural recuperative powers. For all its evident technical backwardness, Medieval medicine would therefore seem to stand in shining contrast to the extreme specialization practiced in a modern hospital, where the human being finds him- or herself often reduced to a faceless conglomeration of clinical symptoms.

school's masters or containing actual summaries of experiences in Salerno. By far the most popular among these often primitive textbooks or household manuals was the *Regimen sanitatis Salernitanum*, a versified anthology of practical experience, probably from the thirteenth century, that, for all its abundant misinformation, may have done more to help people preserve their health during the later Middle Ages than all the learned classical treatises combined. (The emphasis seems always to have been more on the preservation of health rather than the cure of diseases, an emphasis in accord with the concern for the entire person instead of the isolated clinical symptom and its pathology.)

To be sure, such compendia were filled with the most hair-raising misconceptions; these had to be revised against a more carefully controlled experience or slowly replaced by more advanced medical knowledge from the Arab world (to the extent that that had been successfully unraveled and readied for practical applications). Nevertheless, these quaint popular handbooks did their share in disseminating hygienic, medical, and pharmacological knowledge among the public, attracting students to the study of medicine and stimulating others to write more elaborate commentaries — and in this way contributed, by a historic process of hit and miss, to the gradual refinement of the healing arts.

Medieval medicine was very much a combination of native experience and needs, with the Hellenistic and Islamic teachings slowly assimilated; and through this feature the progress of medicine closely resembled the general evolution of Medieval science. A fiercely warlike yet rapidly urbanizing society obviously had no time for acquiring its therapeutic art solely from the ancient textbooks. It had to resort to native resources and pragmatic ingenuity. A surgeon operating on a soldier at the edge of the battlefield, a physician fighting the ravages of the plague in one of the densely populated towns during the Black Death had to use what he knew or what came to hand — and wherever the ancient authorities were able to offer useful help (in the case of epidemics they notoriously did not), he was, presumably, thankful.

The study of medicine at Salerno, which grew out of native experience acquired on native soil, was strongly flavored with international elements in this lively commercial center, where Greek was spoken in the streets; and the translations from the

Arabic were welcomed as an additional, though of course highly appreciated, bonus.

When a rich cache of translations of Arabic medical works (some, like Hippocrates' and Galen's, originally from the Greek) reached Salerno during the eleventh century, it seems characteristic that the cosmopolitan city, at that early date, already had its physicians — clerics as well as laymen — and that the physicians already had students who had flocked there from all over Europe to watch them at work.* If the budding medical school could do little more with the ancient texts than teach them in straight academic fashion as solemn "authorities" whose meaning one struggled to grasp, the Salerno physicians certainly brought a good deal of practical experience to their classrooms, even the occasional use of a pig or a dog for dissection. (The dissection of human bodies was to remain for the future.)

* * *

Medieval Europe's most original contribution to the growth of medicine was perhaps made by the herbals. Medieval society had its compelling reasons for observing the properties of plants and herbs: after all, the basic food supply depended on the intensive cultivation of the soil, including an intimate knowledge of plant life. Supplementing the daily diet through imports would have seemed a fanciful idea — conceivably to be reserved for princes, the sort of high society who might give each other the present of a boar on some stately occasion. Even the use of spices remained a privilege of the rich, with a gradual and broader dissemination from the thirteenth century on.

Meanwhile, monks kept their little kitchen gardens in back of the monasteries or in the center of the cloister, tending the home-grown spices that would flavor the food and make it as tasty as the expensive imports from the East; or growing the herbs that would soothe the pain of some poor devil lying sick on his back in the infirmary. Their findings were carefully recorded in herbals and medical manuals, together with the lion's teeth and the lion's fat and lion's brain and other such nostrums.

* The translator, by the way, at that time a monk at Monte Cassino, was the same Constantine the African who had introduced the West to the Arabic concept of the zero.

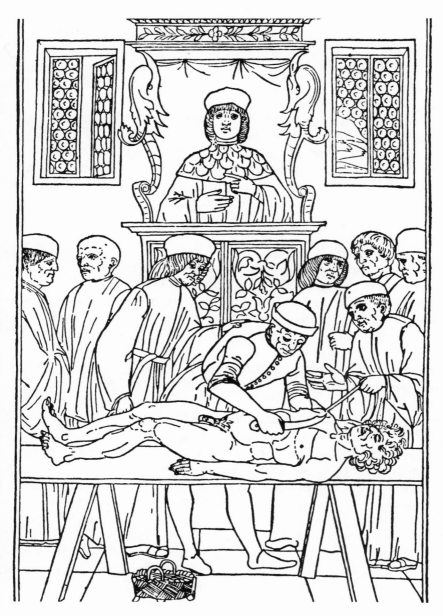

Dissection performed under the supervision of the famous Italian physiologist, Mondino of Luzzi (from a fifteenth century woodcut). Dissection of the human body became accepted by the fourteenth century.

In time, Medieval Europe developed an intimate familiarity with the healing, and sometimes poisonous, properties of plants, which found their way into fairy tales, popular novels, even drama — like that deadly concoction, "of midnight weeds collected," which Hamlet's Player pours into the ears of the king — a kind of popular pharmacology whose rich lore has only in our days yielded to the feats of synthetic chemistry. Mint, hemp, squill, poppy, fennel, aniseed, henbane, castor oil, mandrake, senna, datura were among the herbs taken over from ancient times (some known already in Egypt), to which the Arabs had added their lore. The Middle Ages extended the list and its medicinal applications.

* * *

The observation of plant life follows the general patterns of the mind in its changing relationship toward nature. It reached great heights in Greek and Hellenistic times — in the botanical descriptions of Theophrastus, one of Aristotle's students; or of Crateuas, in the first century B.C. (whose drawings have been partly restored to their original sensitive beauty); or in Dioscorides' famous *De materia medica* from the first century A.D, which became a model for later pharmacopeias and set the basis for botanical nomenclature.

In the botanical sections of Pliny the Elder's *Natural History* the study of plant life is reduced to a hodgepodge, characteristic of the eclecticism of imperial Roman times. Then, with the disintegration of the ancient world, botany seems to have faded into mere decorative imitation — the lifeblood of firsthand observation gone — in the rigid illuminations of early Medieval manuscripts. The early Middle Ages' idea of plants paralleled their ideas of heaven and earth in cosmological thought.

Yet, just as in the cosmic picture, at the very high point of Medieval culture came the revival. Our finding the familiar names among the transmitters of earlier learning, the innovators, the new centers of study, is an indication of the universal appeal of the new study of nature. The same people and places seem to have been involved in all aspects of the new science: Constantine the African (that ubiquitous influence) bringing Islam's botanical knowledge to the Italian South; illuminated manuscripts, serving

An unusually early (and beautiful) example of naturalistic plant studies, a peonia from a twelfth century herbal, written (and illuminated) at the monastery of Bury St. Edmunds.

as models for the herbals of the later Middle Ages, produced at Salerno; the first known actual Medieval herbal, probably composed around 1100 by Odo de Meung, in the form of a poem (the enormously popular *Macer floridus*), which again relied largely on Constantine's information; or a dictionary of drug synonyms, the so-called *Circa instans*, compiled by a physician from southern Italy, one Mathaeus Platearius, who also seems to have practiced at Salerno.

Albertus Magnus took a major step forward in botanical studies with his treatise *On Vegetables and Plants*, in which he mixed data from a pseudo-Aristotelian work with fresh, firsthand descriptions. His observations were remarkable as much for their original insights as for their broad scope. Setting them forth by way of commentary (or *digressiones*) on the presumed Aristotelian text, Albertus presented a massive comparative study of plants, encompassing all the different parts — root, stem, leaf, flower, fruit, bark, and so on — identifying the basic types of floral forms and fruits and recognizing the functions of the sap and the structure of seeds, fruits, and flowers.

But when it came to the healing virtues of plants, the old magic once more reared its head. In the same treatise Albertus respectfully deferred to "the magi" for more detailed knowledge regarding the "divine effects" of certain plants (such as love charms and the hypnotic powers of certain juices). On other medicinal properties he was more explicit: women could protect themselves against pregnancy by wearing a certain herb around the neck; a parsley root dangling from the neck would cure toothaches. To certain plants he ascribed an amazing range of therapeutic effects, as when he wrote of the nasturtium: "It possesses acidity . . . acts as a gentle purgative and laxative, and dries up the putridity of an empty belly. Used as a potion and liniment, it keeps the hair from falling out. Combined with salt and water, it cures abscesses and carbuncles . . . It purifies the lungs and relieves asthma by its sharp, cutting qualities . . . It acts as an aphrodisiac . . . [and] is good for venomous bites."

Albertus attributed the medicinal powers of herbs to the fact that they grow closer to the ground and "recede less from the first fertilizing humor in the earth." Plants, he thought, receive their occult virtues primarily from the movement of the planets at the time when the young plant is formed, because potent vapors ascend from the depths and meet the descending dew.

That Albert could both condemn and accept magic beliefs seems no longer surprising. *"Magnus in magia"* (great in magic), people are supposed to have said of him. What is more noteworthy is his combination of the most penetrating understanding of plant life with a mixture of "earth mysticism" and astrology. One senses, in his feeling for the mystical powers inherent in the soil, his kinship with the alchemists' approach.

Yet to point out his frequent methodological contradictions is, in a way, a pointless game. He earned his epithet, "the Great," because he was a living embodiment of the learning of his time, the thirteenth century, even though he embraces a great many of its foibles. His mind was more encyclopedic than critical; he had a stupendous capacity for absorbing details from the Greek and Islamic literature, plus a genuine pleasure in natural observation. Both gifts sustained him through his fifty years of toil in achieving the great aim of his life, which was nothing less than to restore the whole of Aristotle's natural philosophy for the uses of Medieval Europe.

He succeeded in creating a vast storehouse of naturalist data, drawn both from ancient sources and immediate observation, pervaded by brilliant flashes of original insight. The labor of critical sifting, of devising a valid empirical method, was left for others to refine or complete. His impact on Roger Bacon's methodological intuition and on Thomas Aquinas' philosophical system was crucial. His studies of plant life, regardless of their numerous magic or mystic residues, have ranked him as the first serious botanist since the days of Theophrastus, the founder of scientific botany. His zoological observations were almost equally fundamental.

Tradition portrays the great Albert as a diminutive man of relentless energy, profoundly religious, who kept expanding and revising his writings into his final years, after his retirement to a convent at Cologne, where he died in 1280. He had been born, toward the end of the twelfth century, the son of a south-German nobleman, the Count of Bollstädt; had joined the Dominican order while he was in Italy; had taught at Cologne and other German universities as well as at Paris; and at one time was bishop of Regensburg. He was so pious that it is said he traveled barefoot on his long official inspection tours — while he made his invaluable firsthand botanical and zoological observations — and to have rejected all material possessions, even his own manuscripts.

As an old man, already in his eighties, Albert once more left the quiet of his retreat to defend the memory of Aquinas, who had died three years before and whose teachings were involved in the bitter backlash of 1277. A tiny, fantastically vigorous man, fervidly religious, fiercely loyal to his friends, who almost single-handedly built the foundations of several major empirical sciences from ancient and immediate materials — so the great Albert von Bollstädt stands before the remembering mind. The German nobility has not produced very many giants of his intellectual scope.

* * *

Astrological beliefs might link the action of an herb to the course of a star. Elementary earth magic might envelop its healing virtues. A prayer might be addressed to a saint or the Holy Virgin to re-enforce its therapeutic effect. In one form or another those secret forces of the invisible world might be invoked in the sickroom. Yet for all its mysticism, the Medieval milieu has left its unmistakable imprint on the body of medical lore that the West inherited from its more sophisticated Islamic teachers.

By the middle of the twelfth century the Salerno doctors recommended the use of "anesthetic sponges." * From then on physicians began to experiment with more powerful mixtures, finally including alcohol fumes. A fast-growing botanical literature helped to identify the herbs and plants of the monastic gardens. The number of manuscript copies that can be found in the archives reflects people's increasing interest in the subject. In fact, by the fourteenth century the earliest actual botanical gardens were established, one at Salerno and another in Prague.

In time, the identifications became more accurate, the medicinal applications safer, the illustrations increasingly lifelike. At last, by the fifteenth century, botanical illustrations had reached decided

* Michael Scot, who studied at Salerno, has left us the prescription: equal parts of opium, mandragora, and henbane, to be mixed with water.

Opposite: Use of alcohol for anesthetic purposes in a monastic hospital (from an early sixteenth century Swiss chronicle).

artistic beauty and freshness and developed into a central motif of Renaissance art. In the paintings of Botticelli and Leonardo da Vinci, and in Leonardo's and Dürer's botanical drawings, it is evident that the observations of plant life had become a major creative theme. Indeed, as one looks at one of Leonardo's beautifully explicit flower studies, it is often no longer clear whether one is seeing a botanical illustration or a sample of artistic creativity.

Modern medicine and pharmacology have emerged in outright continuity from the enthusiastic experimentation of the Middle Ages. It was in that mystical setting that the ancient heritage was revived for intensely practical uses. It was then that the first medical school ushered in the rise of universities in Europe. It was at the School of Salerno that medical students, for the first time in the history of the West, were exposed to a challenging combination of ancient learning and contemporary practical experience. More than any other branch of science, Medieval medicine forced its practitioners into a pragmatic fusion of the oddest and most widely divergent elements — dry textbook learning and clinical practice; magic, superstition, and careful empiricism; theoretical analysis and freshly observed dissection — all under the urgent sting of practical and inexorable needs.

The rich lore of the herbals confirms this mixture of outright magic and effective practicality. There is many a medication or pill prescribed by modern doctors within the sober ambience of their offices of whose properties they know little more than the Medieval monk who first discovered them in his tiny kitchen garden — little more than that they relieve the excruciating pain of some poor suffering soul.

* * *

If medicine, pharmacology, and botany were fusing mystical wisdom and practical experience in a baffling concoction, consider the phenomenon of the Gothic cathedrals: Were they the product of scientifically calculated engineering design — or of pure mystical visions, coupled with magical practices of a particularly secret kind?

They were all these things at once, under the same vaulted roof,

One of Leonardo da Vinci's flower studies. Art seems to have fused with the accurate rendering of botanical detail.

within the same complex architectural structure. For one, Gothic cathedrals were miracles of technology. The high pointed vault was the result of bold experiments in the distribution of heavy masses of weight. The builders of the Romanesque churches, preceding the Gothic, had discovered (or rather rediscovered) the phenomenon that the masonry will hold itself in place in an arch or a vault, especially when the structure is reinforced by a framework of "ribs." By the early twelfth century the designers of the first Gothic cathedrals, at Chartres and Paris and places nearby, found the same laws held true when the tip of the vault was pushed upward, to a point considerably above the natural curve of the arch, where the ribs converged and the strains and stresses were concentrated.

When the experiment was multiplied, the resulting patterns of successive or intersecting pointed arches created for the viewer an extremely pleasing esthetic effect, a sense of optical elevation, of being lifted off the ground; the eye was attracted irresistibly to this upward thrust of the ceiling, so eminently suited to the religious purpose. Evidently such a daring structure had to rest on the most precise mathematical calculations, involving a thorough knowledge of the laws of statics. The Gothic architect had to convey the most detailed instructions to his workmen if he wanted to secure the gigantic edifice against eventual collapse.

In fact, the cathedral architects — it was only a passing error of modern historians to think they were anonymous — managed to raise their buildings to ever greater heights, replacing ever larger areas of solid masonry with the sheer force of statics, so that the cathedrals at the peak of the Gothic style stand almost like towering skeletons of curving masonry, with the empty spaces filled by the delicate stained glass of windows or else awesomely bare.

We have enough contemporary records, like the famous *Sketchbooks* of the architect Villard de Honnecourt, to give us an idea of how much painstaking calculation, how much practical knowledge of physics actually went into this kind of design. Although the actual blueprints seem to have been based on an amazingly primitive method, the method proved effective enough to support these miracles of Medieval technology for close to a thousand years. From a purely technical point of view, the experiments of

The main nave of the Cathedral of Chartres shows the esthetic effect of the intersecting Gothic arches.

the Gothic builders were the forerunners of the skyscrapers of the modern world.*

* * *

Still, these experiments were hardly conducted for the sake of pure craftsmanship, in a spirit of cool rationality. During the early phase of the Gothic style, bands of faithful enthusiasts — ordinary layfolk under the guidance of architects or assisted by craftsmen — could be found trekking from site to site, carting the brick and mortar to build another cathedral in the honor of the Holy Virgin or God. Many of the cathedrals of northern France were built by this spontaneous lay movement, the "Gothic crusade." They were built in a great wave of mystic fervor, by young people or grown women and men, who were passing the bricks from hand to hand and chanting hymns to the rhythm of their labor, or intoning the holy songs around their campfires at night.

Nor were professional architects or craftsmen aloof from such impassioned motivations. The truth is, the Gothic cathedral embodies an utterly irrational experiment. The scrupulous know-how that binds the tender filigree of stone with solid physical laws represents a rare union between mystical vision and practical experience. The cathedrals are works of art inspired by visions, not mere buildings, but they are artistic creations in which the technological accomplishment was of the highest degree. Nevertheless, the vision was always the decisive factor.

At first glance, the Gothic structure represents movement. It is a movement that involves every tiny detail, gathering force as it strives upward, culminating in a mighty thrust toward the sky. One may see a Gothic cathedral as something like the representation of the collective religious aspirations of a Medieval town, its spires lifted toward heaven in a gesture of prayer.

Some such tangible symbolism may well have been there, but upward movement in the mystical philosophy had a more explicit spiritual meaning. It signified the individual's striving for God, self-improvement within the religious context. All the steps and

---

* The vault of the cathedral of Beauvais corresponds to the height of a modern fourteen-story building; the spire of Chartres cathedral to a thirty-story skyscraper; the spire of the cathedral of Strasbourg, 466 feet, to a forty-story structure.

stages in the personal process were minutely discussed in the mystical writings. They corresponded to the various levels of the cathedral's structural upward thrust. Documentary evidence leaves us no doubt that this was, in fact, part of the builders' intent.

But it was only a part. The Gothic cathedral was designed not merely to symbolize — and stimulate — a person's self-elevation toward the heavenly universe; it also brought the cosmos down to the physical level of the Medieval city. The cathedral symbolized humanity's mystical union with the Divine — and accomplished that union from either end.

A forceful reminder that the Christian stands in need of such union was urgently called for by the time the cathedrals were built. The Gothic style was developed in the early capitalistic towns, where trade and primitive industry were producing the beginnings — and fostering some of the mentality — of the modern urban environment inside their Medieval walls. The cathedrals were the stony embodiment of a message the Church was preaching to the new urban masses. The message urged the townsfolk to look up from their petty material concerns and remember that the true life is in the beyond.

The cathedral represents that beyond. True, the exterior tends to emphasize spiritual striving. Once inside, even the modern visitors can have little doubt that they have stepped into another world. The high vaults, awe-inspiring, distantly echoing the human voices, and the great solemn dimness are the closest attempt ever made to involve us in a representation of the heavenly universe.

Still, the majesty of the universe as a purely abstract idea cannot be expressed in visible terms. The cathedral represented a concept of the universe that was natural at the same time that it was divine. Religious reverence united itself with the kind of awe we may feel before the vastness of the stellar cosmos on a starry night. The cathedral, with its complex and subtly planned design, was conceived as a "mirror of nature," of the natural universe as a whole, which itself was understood as the work of the Heavenly Architect.

It was the intrinsic lawfulness of the cosmos, the ultimate unity of divine and natural law, that the cathedral builders meant to evoke. Only when the lawfulness of nature had been recognized as a divine manifestation — as it was at Chartres* — could people conceive of creating a structure that would symbolize the natural universe in its full majestic solemnity. The fact that nature proceeds according to certain invariable laws (a fact that dawned on the Middle Ages with the force of an elementary discovery) did not then seem self-evident. Rather, it was a particularly persuasive expression of God's wisdom.

This is, of course, a mystical, or at any rate metaphysical, concept, born from an essentially religious feeling, and its relation to nature rests in a kind of "nature mystique" more than anything else. But it was close enough to the actual scientific thought of the age to permit a union of the mystical features with mathematical calculations based on clearly perceived physical laws. The Gothic builders could draw on both facets in realizing their vision.

* * *

The message was strengthened by the use of light. In the catalogue of the mystics, sunlight had an especially elevating meaning. Light was uplifting, exalting. To open oneself to the rays of the sun meant to cleanse the dingy recesses of the soul, baring one's soul to the touch of the Divine. Medieval people seem to have been attuned to light, as we are to music. As a few well-harmonized chords may transport us in an instant away from our trivial cares, filling our souls with a more fundamental and sublime meaning, so the beams of sunlight slanting through a choir window into a cathedral could link the devout at once with the universe. Stained-glass windows high up in the side walls might add variations to the basic theme, by filtering the light through patterns of poignant blues, reds, greens, and other colors, making them sing out like voices in an angelic chorus.

* * *

The art of the stained-glass windows had developed apace with the Gothic style. The sunlight streaming in from behind the altar was consciously used by the Gothic builders to replace somber darkness with gleaming light, spreading upward into an evocative dimness.

* Or, later and quite explicitly, in Thomas Aquinas' philosophy, where the idea is central to his thought.

The rose window at Chartres's south portal is a beautiful example of the use of light in the Gothic cathedral, filtered through the multicolored stained glass of the windows.

When the Abbot Suger was revamping the gloomy interior of the Church of St. Denis, north of Paris, turning it into one of the earliest examples of the new style, he made sure to surround the choir with large windows so that the heavenly light might enhance the beauty of his handiwork. An inscription proclaimed his pride in the result:

> Once the new apse is joined with the old façade,
> The center of the sanctuary gleams in its splendor.
> What has been splendidly united shines in splendor
> And the magnificent work, inundated with a new light, shines.
> It is I, Suger, who in my day enlarged this edifice,
> Under my direction it was done.

Light was deliberately used as a structural feature of the new style.

Sometimes the builder might feel he was carrying out some explicit heavenly command. Many of the early cathedrals were built in honor, and under the presumably immediate guidance of the Virgin Mary.* The saints, speaking through the language of the miracle, might instruct a builder to decorate some interior detail in a particularly lavish way. Explaining his use of sumptuous decoration for an altar in front of St. Denis's tomb, the Abbot Suger cited express heavenly orders: "While we, overcome by timidity, had planned to set up in front of this [altar] a panel golden but modest, the Holy Martyrs themselves handed to us such a wealth of gold and most precious gems — unexpected and hardly to be found among kings — as though they were telling us with their own lips: 'Whether thou wantst it or not, we want it of the best.'"

In our modern skepticism we are convinced that the good abbot was merely a clever promoter who knew how to justify his extravagance before his more ascetic contemporaries. Perhaps he was. Yet he may just as well have believed that he was in touch with the spiritual world of the beyond, because that was what his culture insistently taught him. Besides, who but someone completely filled with the presence of the spiritual world could have devised the fundamentals of the Gothic cathedral?

The sense of communicating with a universe that was at once

* Notre-Dame of Chartres, Notre-Dame de Paris, of Laon, of Amiens, and others have preserved the origin in their names.

natural and supernatural — the true universe of the mystic — might cause the builders to overrun the cathedrals with representations of the natural world that were often conceived in fantastic, at other times in perfectly realistic, terms. Yet always there is the inclination of the mystic to see the objects of nature as symbols, full of poetic or spiritual meaning. The strange creatures captured in stone might mirror a fascination with the wonders of distant lands (still unchecked by direct observation): doglike beasts with human feet; human figures with horses' hooves; monsters with human faces; the dreaded griffin, part lion, part eagle, who supposedly guarded the treasures of the world; or gargoyles that looked like mixtures of human beings and giant bats.

Most of these monsters had sprung from the pages of the early Medieval encyclopedias, such as those of Isidore of Seville or Honorius of Autun, or the famous *Polyhistor* from the third century. But what the encyclopedists had done (and the Gothic sculptors had copied) was to people the unknown parts of the earth with the grotesque products of a rampant unconscious. The mystical vision was blurring the boundaries toward the supernatural by distorting the natural reality. The products of the Gothic imagination were hovering in a twilight zone where reality became fantastic, and the imaginary world reality.

For all that, a good part of the sculptured decoration that flourishes among the pillars of Notre-Dame is simply a stylized version of the humble plants that grow in the woods and meadows of the Ile-de-France — fern, clover, buttercups, snapdragon, strawberries, parsley, cress, oak, and other samples of the local flora. Art historians believe they have discovered a curious fact about these homely decorations: the early Gothic churches show a preference for the flora of early spring — young shoots and buds, curled-up young leaves bursting with sap. As the style came of age, in the course of the thirteenth century, the preference shifted to a more advanced season of the year — the buds tended to burst, the leaves opened up — until, a little later, the doorways and pillars might be wreathed with vine shoots or rose stalks and fully grown branches or even, toward the decline of the Gothic style, with the autumn thistle. The evolution of cathedral decoration seems to have followed the natural rhythm of the seasons.

The phenomenon implies an unexpected closeness to nature on

Gargoyle at the Cathedral of Notre Dame, Paris.

the part of the Gothic builders, to a point where the poetic symbolism of natural growth must have entered into their thoughts almost unconsciously. The scents and saps of buds and flowers, their images popping from doorways and capitals, the natural progression of the year, the wildlife of monstrous exotic beasts — which were, of course, imagined as being real — cowering atop a flying buttress or chasing each other around some pillar base, filled the cathedral with a breath of natural life that extended the scope of the heavenly universe to embrace all our earthly nature.

*　*　*

For a long time it was assumed that still another mystic or, rather, magic dimension was added by the ordinary workmen crews under whose hands the huge structures arose. By now, under the sharp eye of critical scholarship, the case looks more doubtful.

One way or another, the evidence rests largely in the mysterious marks the masons left on the surface of building stones all across the Gothic structures. What was the meaning of these symbols? Did they denote some secret invocation of supernatural forces, magic symbols each generation of craftsmen left for its successors, comparable to the magic formulae of the alchemists? Did they merely serve a functional purpose, such as position marks, to instruct the masons which stones to place next to, or on top of, which? Or were they some abbreviated form of the workmen's signatures, from which the foreman might compute a mason's piecework wages at the end of the week?

Scholars have tried to compile careful panels of masons' marks from the different cathedrals and concluded that the functional purposes would fully suffice as a key. One symbol might be the initial of a mason-father; the same symbol with a slight variation that of his son. Another type of incision served doubtless as a position mark. And therewith the case seems to be closed, the magic banished at least from this down-to-earth, work-a-day sphere.

But is it? A touch of mystery continues to surround the masons' craft. Migrant workers who, for the most part, journeyed from town to town wherever some important construction project was under way, they were organized in a type of guild or workshop — "Bauhütte," "chantier" — which, because of the transient life

Examples of stonecutters' marks, from Strasbourg Cathedral.

style of the workmen, evaded the sternly regulated life of the city guilds and was often a center of political unrest. In time these nuclei of dissent began to attract the local intellectuals, who may have provided free lectures for the workmen, participated in their debates, and helped to make the freemasons' lodges a thorn in the flesh of local Church and city authorities. (The paranoiac fury with which modern fascist governments have persecuted the specter of Freemasonry was clearly an echo of the resentment of the Medieval establishment.)

To this day the Masonic Lodges, which developed, in fact, out of the cathedral workshops, are noted for their commitment to the welfare of their members on a far-flung, international scale, a climate of free thought, especially in religious matters, and a distinct propensity for secret ceremonies and rites. Since the first two features clearly go back to their Medieval ancestry, it is likely that the aura of secrecy — the set rituals involving the initiation of new members and the election and functioning of the officers of the lodge — are also part of their Medieval heritage. No doubt the original freemasons' workshops were given to all manner of secret devices (the very suspicions of the local authorities give sufficient grounds for that supposition). Secret rites, secret codes and symbols must have flourished richly.

What is more, Medieval symbols quite often had a dual, even multiple meaning. Not only poems like Dante's *Divine Comedy*, but almost every single manifestation of Medieval art is notoriously pregnant with symbols involving two or three, sometimes as many as five, different meanings. Although modern scholars are struggling to reconstruct them all, they were apparently perfectly self-evident to a Medieval audience.

The rich tableaux of marks the masons left in the face of the stones from which the cathedrals were built may well have had an equally multiple meaning, denoting some sort of magical incantation besides their downright functional significance. Scientific precision and meticulous workmanship went hand in hand with supernatural feelings all across the Gothic constructions. It would not be surprising to find the cathedral shot through with that combination, from the initial conception to the last workmanlike detail.

* * *

The mystical mind worked with symbols and multiple meanings. The mystical imagination could people the world with fantastic creatures, ascribe mysterious powers to the earth and the stars, or include all nature in a religious vision of order and lawfulness. Yet always the mystical vision seemed to perceive something in back of the tangible world, some hidden substance entailing a vast range of "insensible" phenomena — as the Middle Ages liked to call them — things that could not be verified by observation but were nevertheless known to exist.

For us, the crucial question is, of course, whether they actually did exist or were mere figments of a bizarre collective imagination. Yet even if we grant that this whole, largely unverifiable dimension might, in fact, have possessed some degree of reality of which, for instance, modern experiments in extrasensory perception or modern psychology's explorations of the unconscious may be receiving a fleeting glimpse, there is still another way of looking at the problem. It has to do with that gradual change of focus, from otherworldly, spiritual, or metaphysical matters to the clearly identifiable objects of this world, that is common to all these developments.

Clearly, this adjustment of vision was the most important experience of the Western mind in its emergence from the Middle Ages — possibly in its entire history. Neither the brilliant productivity of the Renaissance nor modern civilization would have been conceivable unless certain circumstances had forced the Medieval mind to come down from its lofty heights and embrace the earth, with all its sweetness and sorrow. Nearly every intellectual or cultural development during the transition toward the modern world has to be seen in this context.

Science, in this profoundly unsettling (and profoundly stimulating) process, was cast in a particularly strategic role. Increasingly, the "scientific approach" came to exemplify the modern outlook, focused on the concrete objects of the world, to the exclusion of everything that is not observable with senses or demonstrable through impeccable rational argument.

Obviously, these tendencies have been gaining in power in the course of modern history. More and more, in the modern world, it has become a matter of personal pride to act — at times even to live — according to "scientific" criteria, to think and form one's values in accordance with rational and empirical observation, and to exclude as "unscientific" — or even unreal — everything else.

Paradoxically (but not without a certain inner logic), the evolution of science itself has become imprinted with this process. Those angry, inconsistent rumblings of some thirteenth-century scientists, like Albertus Magnus, against the "false magic" of their contemporary astrologers were, in effect, the opening of a ruthless purge of all "contentious learning," all and any beliefs not tested by downright empirical observation, to which science has subjected itself over the past seven hundred years. The question as to whether we could gain insights into the works of nature by any other approach was brushed aside with the ruthlessness of newly acquired convictions. Whatever did not lend itself to the modern methods of verification and scrutiny was banished from the realms of science and branded with the stigma of fabrication or idle fantasy.

The new scientific spirit entered the early modern age with assertive self-confidence, stemming from its methodological assurance.* It was cultivated in the emphatically empirical climate of the School of Padua during the fourteenth century; triumphed in the stalwart pragmatism of fifteenth-century geographers; decisively informed all of Galileo's writings; and finally found its classical — its most explicit and most aggressive — formulation in the writings of Francis Bacon, early in the seventeenth century.

Although modern science has profited incalculably from this self-imposed limitation of the focus of research (and the attendant methodological clarifications), it is evident that such progress was achieved at a heavy price. Precision was gained at the expense of an intrinsic largeness of the view; the elimination of all that is unverifiable (and of much that was no more than fantasy) at the expense of a sensitive open-mindedness toward psychic phenom-

---

* Methodological skill (as well as the characteristic emphasis on the scientific method) was in essence developed as a response to the Medieval West's encounter with the overwhelming mass of empirical data presented by Islam. Medieval scientists were also stimulated by certain methodological innovations in Islamic optics, chemistry, and medicine. The first major — and seminal — exposition of the scientific method in the West was made by Robert Grosseteste (circa 1168–1253), at one time chancellor of Oxford University, who established, and thoroughly discussed, its three principal aspects, the inductive, the experimental, and the mathematical. (See A. C. Crombie, *Robert Grosseteste and the Origins of Experimental Science, 1100–1700*, Oxford, 1953.)

Narrow street in Siena.

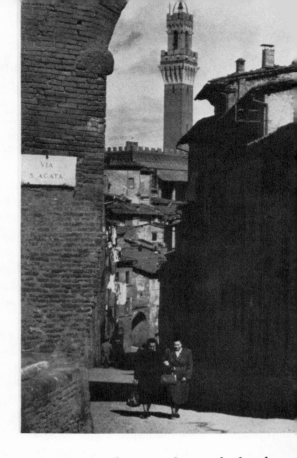

ena — or toward a realm of existence that may be reached only by unhampered speculative thought, sometimes merely by a kind of groping intuition.

* * *

On the road from Florence to Rome lies the town of Siena. Nowadays, the bus — one of those streamlined Italian excursion buses, complete with attractive multilingual guide — spirals its way toward the center of the town, past a series of suburbs scattered over outlying hills, as if peeling the layers off an onion before reaching the core. The core is the Cathedral Square. Nearby is the large shell-shaped square of the Campo, with the lovely Palazzo Pubblico, the fourteenth-century city hall.

The town — narrow streets climbing up and down over worn steps and bumpy cobblestones between tall Medieval buildings, gray houses topped by weathered orange-tiled roofs — stands virtually unchanged since the fourteenth century. Nevertheless, it is bubbling with life. To be in Siena is to get as close to the life of a Medieval town as one can anywhere today.

Life in Medieval Siena had certain definite modern elements. The city did well in the fourteenth century, partly by selling produce from the surrounding countryside (as it still does), marketing wine, vegetables, and olive oil, playing host to the tired and hungry tradesmen who were passing through over the highway from Rome to Florence; partly by sharing directly in the burgeoning network of international trade and the related banking operations. So vigorous was its native energy that even after the plague had murdered more than half the population, town life bounced back, unfinished building projects were picked up again and completed, and the arts resumed their flourishing state, clearly pointing the way toward the Florentine Renaissance.

Civic pride, that forerunner of modern nationalism, was the creative hand that shaped fourteenth-century Siena (as it stamped the character of much of Sienese art) — the pride of the citizens in their energetic commonwealth. Among the narrow streets of the inner city stands the Loggia dei Mercanti, the Businessmen's Hall, open on three sides toward the street, in the Italian manner, where the merchants transacted their daily affairs. What could be more modern in a Medieval setting, more secular than a city working its way out of the Middle Ages with the help of banking and trade, propelled by its own vital energies?

Sienese art has captured that modern spirit. A mural inside the City Hall shows the fervor of construction work in progress, including the use of advanced technical gadgets, such as a pulley moved by a crank lifting the materials in place at the top of a city gate.* Sienese artists were among the first to discard the bland backgrounds of Medieval paintings and open the view to actual scenery: Ambrogio Lorenzetti showing a city by the sea, or the houses of Siena surmounted by the belfry of the cathedral; Simone Martini depicting a proud knight on horseback riding under a blazing blue sky, among castlelike fortifications; or, again, an idyllic garden scene. In back of the marvelous and imposing cathedral stand

* See photograph in Introduction.

View of Siena with cathedral and Palazzo Pubblico. The remnants of the planned "super-cathedral" are behind the bell tower.

the remnants of an even more ambitious design, a mighty super-cathedral, in which the present structure would have served simply as a nave. Ambitious projects, advanced technological devices, unprecedented vistas of the real world — such is the handwriting of the secular spirit of Medieval Siena.

Yet to see Siena only in this modern light is highly misleading. It is but one aspect of the spirit of this Medieval town. The other is mysticism, which kept its hold on the Sienese mind right through all this ebullient secularity.

To this day, the people of Siena regard as their most precious painting a work by their painter Duccio, the *Maestà*, or *Majesty*, meaning the supreme majesty of the Queen of Heaven, who sits on her throne, the child Jesus in her lap, her head slightly inclined, surrounded by saints and angels. Originally painted for the high altar of the cathedral, it is now the showpiece of the Cathedral Museum, where, in a small, dimly lit room, it occupies an entire wall. Suddenly, after contemplating it for a few minutes, one realizes that one is looking not at just another painting of the Virgin, not at a marvelous rendering of a familiar religious theme done in exquisite colors in the late Gothic style. One is looking into another world.

Not one of the products of Medieval culture can be properly understood unless one allows one's senses to ascend to this region. The tall buildings and superstructures, the bustling merchant exchanges, the murals, with their glimpses of landscape or cityscape, the technical devices, even those closely reasoned theoretical speculations of the scientists, only seem to be modern. The mind that conceived them still dwelled in the unchanging calmness of the world of Duccio's *Majesty*.

Its most distinctive feature was not that it was a religious world. It was more. It was a source and center of thought, a quiet pole to which the mind could return at any moment from all the earthly turbulence, a calm assurance that there are forces beyond the noisy confusion of this world that are the timeless movers of life and of ordinary fate. The people of the Middle Ages would indeed approach this region through the customary paths of religion. But this transcendental dimension could as well appeal to the senses, just as Duccio's *Majesty* seems to transport us into another sphere with the gentle force of a sequence of musical chords.

Central portion of Duccio's MAESTÀ (1310), originally painted for the high altar of Siena's cathedral, now in the Cathedral Museum.

It was there. It existed. Its presence was taken for granted — not like the physical trappings of modern urban living to which we no longer pay any heed, but as a continuously felt presence that subdued and tempered the business of the everyday world. It generated its own quiet music (after all, painting was only one medium, humbly considered inadequate, by which to capture that mystic reality). The life of smaller Italian towns is still steeped in its rhythm, a rather remarkable phenomenon: the door of a shop opening or closing, setting off the tinkling of a tiny bell; the voice of a vendor calling out wares, vanishing amidst the labyrinth of the narrow streets; the brooding of the *siesta* hour; footsteps up or down worn steps; the rhythmic pumping of water from an open well and the clatter of copper vessels; the cries of children at play in the distance; the sudden burst of laughter from the open vault of an inn; and again the silent presence — the great, meaningful, living, breathing presence, never empty or oppressive — that seems somehow caught between the walls of the ancient buildings (or suspended in the air) but is really only the rhythm to which the town had adjusted many centuries ago, like a silent theme around which all the everyday noises have reverently arranged themselves.

It is from such living residues, not from any abstract cerebrations, that we must seek to recapture something of mysticism's creative force.

* * *

The idea of a living substance — known to be there, though concealed from the senses — is at the core of Medieval alchemy. Alchemy was essentially a system aimed at the materialization of that same "insensible" substance and its utilization in the service of the human community. Alchemy was not simply a fraud perpetrated on gullible townsfolk, nor sheer nonsense, nor even a crude get-rich-quick scheme to fabricate gold. It was an incredibly patient, systematic attempt to conjure up that living substance in tangible form and use it as a kind of amorphous raw material for the creation of utterly unheard-of new elements.

Whether we choose to think of it as a science, pseudoscience, or "art," alchemy certainly attracted its portion of charlatans, its mercenary "gold-makers," its share of crackpots. For its serious adepts it was a science, or, as every Medieval science was called, an art.

From a sure faith in the reality of that living presence, alchemists were trying to prod, force, or seduce the unknown gigantic forces into assuming tangible shape. In a general sense the centuries of alchemical experimentation anticipated some of the most crucial modern discoveries, from chemistry to electro- and biochemistry, to electro- and nuclear physics.

Unknown for thousands of years, these forces existed. Nature was alive with gigantic powers waiting to be freed from their slumber and put to practical use. All that the alchemists were lacking was the proper method or "formula" — so they believed. But what they contributed was an exhaustive testing of every conceivable alternative approach, till only the proper solutions were left.

*   *   *

We can visualize a Medieval alchemist at his work in one of the oldest quarters of Paris, like the one formed by the narrow and noisy streets in back of Les Halles, the sprawling old indoor market that seemed forever drenched in the smells of fruits, vegetables, and fish, with only an occasional whiff from the flower market to scent the air (at last moved, 834 years after its founding, to more sanitary, less colorful quarters). A short street off the Rue de Rivoli, not far from the Rue des Halles, was named after Nicholas Flamel, a famous alchemist, who lived and worked there during the fourteenth century (and whose memory has been the object of some controversy and doubt among later scholars).*

The inside of an alchemist's shop presented a rather fantastic mélange between a scientist's lab and a magician's den. In a way, a man like Flamel was both: practitioner of a scientific tradition going back to ancient India and China (very likely even to prehistoric times), yet by the same token adept of the Hermetic arts, a

---

* Other European cities have their "alchemists' streets," like the Gold Maker's Lane, Zlatá Ulička, a tiny side street off the Hradčany, the magnificent old castle of Prague — a row of small, gnarled, twisted houses that look like little old men cowering under their shingled roofs, topped by the hewn chimneys that were an inevitable adjunct of an alchemist's shop. It is hard to imagine that these charming architectural oddities, their quaint shapes an apt reflection of the bizarre experiments that went on inside, were the forerunners of the cold glass-and-steel structures housing modern laboratories.

ZLATÁ ULIČKA ("Gold Maker's Lane"), Prague.

time-honored tradition of magic.* The two strains, the body of factual experience and the magic practices and rites, had become inextricably intertwined across the ages, forming one of the strangest subcultures in history. Secret metallurgical techniques, mystical ideas about the origins of minerals and fire had blended with the primeval cult of the "Great Earth Mother." Early cultures had believed that the uses of fire were divulged by the gods; that

---

* The term "Hermetic" was derived from Hermes Trismegistus, the Greek name for the Egyptian god Toth, who was supposed to have written a metaphysical work on alchemy, magic, and astrology — no trace of which, of course, was ever found. The Greek "Trismegistus" means "three-times great."

the Great Mother goddess had yielded her bowels to the extraction of minerals and metals.

In the shop of the fourteenth-century alchemist these ancient myths, luxuriantly overgrown by the ages, lived on. In his back street in the old quarter of Paris, ancient volumes and strangely shaped phials looking down from the shelves, Flamel cultivated what he could regard as true science but was, in fact, still imbued with the rituals of some five thousand years past. We may call it a pseudoscience; its practitioners dilettantes. Contemporaries suspected the claws of the devil. But the satisfactions inherent in the alchemical experiments, the careful transmission of knowledge, the probing into a secret world that seemed real enough, though reluctant to surrender its secrets — such were, no doubt, the satisfactions of the scientist and scholar. Beyond some explicit techniques — and some rather astonishing insights — Medieval alchemy has taught modern science the value of a thorough grounding in patient and systematic research.

*　*　*

The saturated light of a Paris afternoon — that singular atmosphere that brings out the colors as if they had been dappled over the objects fresh from a painter's palette — falls on an assortment of the quaintest utensils. Glass vessels and vessels of tin, copper, or earthenware; gourds, mortars, crucibles, flacons and pots of orpiment; bellows, tripods, strangely shaped test tubes, and primitive heaters are crammed into the shelves, nailed to the wall, or strewn across the floor of the workshop. The principal fixture in all this confusion is the athanor, the alchemical furnace. (The word, from the Arabic *al-tannur*, recalls that Islam had once more been the link to the ancient tradition.)

Here is the center of the alchemist's activities. Equipped with a glass mask, armed with his bellows and retorts, shuttling between a still (the alembic) and other assorted pieces of apparatus, it is at the furnace that he puts his experiments to the ultimate test. He has pored long hours over volumes containing strange formulae, written in stranger symbols, for the substances to mix, the degree and type of heat to apply in the furnace. Metals were liquefied over the fire; other ingredients added. At last he pours the finished

An alchemist's shop with various characteristic instruments, including the al-tannur.

concoction into an egg-shaped hermetic vessel (said to have been invented by Hermes Trismegistus), a forerunner of the modern laboratory retort (called "phial," "nuptial chamber," sometimes "philosophical egg"), and heats it over the athanor for the *magnum opus*, the ultimate feat. As he steps back, as the weird mixture escapes in smoke through the chimney, or perchance explodes, shattering the equipment — exactly what is it he has hoped to achieve?

The idea that an alchemist was out to "make gold" goes back to a peculiar Medieval fallacy. Casting suspicious glances at the chimney-crowned alchemists' shops (or scared out of their wits by the periodic explosions), the neighbors decided that what the "puffers" were after must be gold. The proverbial pot of gold seemed the only conceivable prize for such unflagging efforts, involving such evident risks. There are two pat explanations for someone's laboring over an incomprehensible chore: he is either playing for the highest stakes, or he must be a crackpot. In the case of the alchemists, their neighbors generously assumed both.

Gold had a limited, well-defined function within the alchemical scheme. Although there were "gold-makers" in the profession, authentic alchemists were pursuing more serious aims.* In the inquiry after "principles" and "knowledge," true alchemists — representing the substantially scientific tradition, as against the benighted dabblers in magic — considered both gold and the refining of metals as no more than a means to an end; but in their metallurgical experiments the appearance of gold would have been the unequivocal signal that a metal had been refined down to its original purity.

Taken on its literal premise — that gold forms the basic substratum beneath the various alloys of which all metals were thought to consist — the idea was clearly preposterous. Still, after our more enlightened centuries have enjoyed their good laugh, at least one aspect of the alchemical premise has, surprisingly, proven correct: gold can indeed be produced by a chemical process (though through a synthetic, rather than an analytical, operation).

As for the more basic aspects, even here there was a method to their madness, a valid concept behind their obsessive pursuits. It contained the seeds of the most fertile innovations by which the modern West has outstripped the scientific efforts of all earlier civilizations. It was a matter of "method," an intuitive feeling for a strange new approach, one that might unleash nature's invisible forces. Disguised in a mystic garb, conceived as something like a magic wand, this intuitive notion was inherited by the Middle Ages from the Islamic alchemists, who had, in turn, learned it from the older cultures. A few brilliant men, like the Arab Geber,

---

* "False alchemists seek only to make gold; true philosophers desire only knowledge," as one of the serious researchers put it, adding: "The former produce mere tinctures, sophistries, ineptitudes; the latter inquire after the principles of things."

---

may have foreseen the actual scientific possibilities as early as the eighth century. But in terms of direct historical causation, it was the Medieval alchemist whose dogged persistence opened the way for the modern approach.

The idea presented itself to him under the concept of "transmutation." Proceeding from the assumption of a primary raw material hidden somewhere in nature that, once discovered, could be used to rebuild the world in purer, incomparably more beneficial forms, the alchemist was in fact working with a remarkable idea involving a wide range of radically innovative techniques. Rummaging among the available elements for that elusive "prime matter," trying every conceivable trick to reduce a metal to some sort of more basic ore, he refined or developed such promising methods as calcination, reduction, evaporation, sublimation, crystallization, and melting. Bent over his bizarre furnaces and stills, he was, in truth, experimenting with the techniques of chemical analysis.

The future has borne him out. It has shown that through a process of analyzing, separating, or "breaking down" the known substances, new, hidden substances can be made to appear, containing undreamed-of aids for the physician, the dietician, the pharmacist, the biochemist, the physical chemist, the engineer. The alchemical notions may have been simplistic, in some ways misguided, the goals scurrilous, often naïve. The general conceptual premise was brilliant. Its practical applications were to bear fruit far beyond the limits of chemistry.

While their explicit procedures developed into the techniques of the modern chemical laboratory, the underlying philosophical concept implied nothing short of a revolutionary approach to the whole natural world — a creative approach reaching beyond the cautiously descriptive boundaries of the empirical method (which can, after all, work only with what is already there) into a hidden and virtually limitless region. Its thrust was toward the "principles of things," as the alchemist said. What the spade or the plow had done for agricultural progress by introducing the tilling of the soil (compared with the mere reaping of the surface crop with a primitive sickle), the alchemists did by piercing the visible surface of the empirical world through the use of the analytical method.

Their techniques heralded the rise of modern chemistry; the analytical principle was to revolutionize the whole range of science and to multiply its creative scope.

*   *   *

How did the alchemist hit on this approach? A certain facility in analytical thought was simply endemic to Medieval philosophy, at any rate in the abstract, rigidly logical form we associate with the Scholastic method. But the most potent incentive grew out of his mystical beliefs. His search for the "principles" of all visible being — the ultimate raw material, the primary metal, or, in a somewhat different context, the elusive "philosopher's stone," entailing the basic principle of the catalyst — was a foray into the unknown, invisible, "insensible" region, the core or pivot of the mystical faith.

To think that matter consists merely of matter — and, therefore, that the analysis of a substance might net further material substances, which are simply "contained" in the first like the progressively smaller dolls enclosed inside a Russian doll — is an assumption peculiar to our modern pragmatic (or "materialistic") view. To the mystic, this simple, and ultimately unphilosophical, premise was in no way self-evident. To him, it was axiomatic that the unknown was more than mere matter. It resided in its own separate region, was subject to its own mysterious laws, possessed its own distinct identity, and could neither be conquered piecemeal-fashion nor demonstrated to be part of the material reality. The point is not which of these views may be philosophically more accurate (at best an unanswerable question). The point is that it was the mystic view of the world, whether "accurate" or not, that in the end managed to summon new giant forces out of the unknown, for the use — or misuse — of modern humanity.

Tinkering with his metals and mixtures, pausing to reflect and review his approach while trying to get at some underlying essence, was the alchemist's way of reaching through to that infinite beyond.

*   *   *

Tangible materials were no more than the medium causing that immaterial world to "appear." Using his own terminology, the alchemist was out to reach the "quintessence" in back of the material world — that very quintessence which, according to Aristotelian physics, formed the substance of the stellar configurations, the mysterious fifth essence beyond the four known elements (or essences) of the earth.* He was using material things as a kind of handle or lever in order to touch that ultimate reality. To his mind, it was his true mission to "free" that insensible substance from its immaterial state, even to "resurrect" or "redeem" it (the alchemists liked to use theological language) from its death-like sleep — thus to succeed the Good Lord Himself in the act of creation.

For all its irrational premises, the scheme seems consistent. As one translates the mystical concepts into the corresponding modern terms, it reveals its remarkable validity. The invisible substance, the elusive quintessence, becomes the site of potential materials or forces heretofore unknown to the human race (even though for the mystic it stood for a great deal more). In place of the one primary matter the alchemist was attempting to conjure up, that unknown dimension proved to be a virtually unlimited arsenal of elements and energies hidden from human knowledge

* The concept of the quintessence represents an especially characteristic fusion of mystical and rational elements in Medieval thought. While alchemists, astrologers, Gothic architects and craftsmen, and others had their own, highly intuitive ideas about that unknown region as the site of mysterious powers, it occupied at the same time a very definite place in the accepted Aristotelian — and perfectly rational — system of physics. Based on immediate observation, Aristotle had postulated a common-sense order of the physical cosmos in which, as he writes in book IV, chapter 5 of his *Physics*, "The earth rests inside the water, the water inside the air, [the air again] inside the ether, and the ether inside the sky, but the sky is no longer [contained] inside anything else." (One can easily visualize this system of concentric shells — or "spheres" — when looking at the ocean from a deserted beach, with the sky as the final, domelike covering. Aristotelian physics assumed that the element "earth" protrudes above the surrounding sphere of water in several areas, forming the continents.)

By a series of astute and rational (if obviously incorrect) inferences, Aristotle had ascribed to that "celestial region," in which the stars appear to orbit around the earth, a number of physical properties (such as weightlessness) amounting to a special set of "celestial" physical laws, as distinguished from the laws of "terrestrial" physics. By the fourteenth century, advanced physical and mathematical thought was beginning to pierce that distinction, extending the laws of motion observable on earth to the celestial region, which led to the eventual discarding of the "quintessence" concept of a separate element. (Some of these advances are discussed in Chapter Six, in connection with the changing concept of space in art.) The point here is that the alchemists' intuitions about an unknown substance beyond the terrestrial four elements were perfectly compatible with accepted physical theory.

till then. As we substitute steam, electricity, or nuclear energy for the vague mystical ideas — the very sources of power for which Medieval craftsmen were groping in tireless searches — the shape of modern technology begins to emerge against the mystical horizon.

Even the concept of a single primary substance underlying the bewildering variety of the known substances seems at least a logical construct. Modern chemistry was still hunting for such a prime substance (called "phlogiston") well into the eighteenth century. Nor were the deeper implications entirely abandoned even by the twentieth. Albert Einstein, especially during his final years, was engaged in a search for some ultimate source of all physical forces under what he termed his "unified field theory." Working with the keenest rational precision, employing the most advanced methods of modern mathematics and physics, he was in essence searching for that life force whose vision had haunted Medieval alchemists.

What Flamel and his colleagues were after was ultimately based on an immensely fertile vision that transcended the limits of modern chemistry. Through the systematic analysis of known matter, we may indeed reach into an unknown region, touching hidden layers of the natural process — and thereby not only expand our theoretical horizons but mobilize those secret practical forces whose presence the mystic intuitively sensed. From chemical analysis to atomic transmutation and nuclear fission; from the chemical test- and distilling tube (or the wide range of agents used as catalysts) to the atom smasher and nuclear reactor, modern science has richly vindicated the alchemical quest. The sharp-edged spade of the analytical method is forever being driven into deeper and deeper layers of the unknown. However seminally or dimly, the full scope of the analytical thrust of modern science, with its steadily multiplying results, was already present within the alchemists' minds.

Of course, something very much akin to a "distillation" or "purification" had to take place to transform alchemist intuition into scientific method — a difficult process that was to occupy a good five hundred years. But should one not credit the creative chaos with having spawned the first seeds?

Besides, alchemists may not always have been as naïve about

what they were doing as it would seem. Roger Bacon, their most brilliant exponent, seems to have foreseen the complete modern scientific method and its potential as early as the thirteenth century, when he insisted that straight empirical observation needed to be coupled with the probing thrust of mathematical analysis. His insistence earned him the dire suspicions of his more sober-minded colleagues. Intuition is not always rewarded. Even the mystic Medieval mind was beginning to feel the first shadows of the suspicious pseudorational pedantry that has often encumbered the positivism of modern scientific thought.

SIX

Art and Science in the Renaissance

As ONE STANDS on a hill overlooking Florence, the perfect Renaissance city seems to extend at one's feet. The town itself looks like the most perfect Renaissance creation — as in fact it was intended and designed: the sweeping panorama capped by the cupola of the cathedral; the Ponte Vecchio, the most spectacular of the bridges across the Arno River, linked with the nearby structure of the Uffizi by a delicate arcade; the whole town with its churches strung along the river like a marvelous, masterfully designed set of pearls. Even the silence reaching across the limpid atmosphere toward the hills, occasionally punctured by the muted sound of a vehicle from the town or the Mediterranean cadences of a woman's voice close by singing while doing her laundry, seems to complete the impression of a tableau, of a city planned as a work of art.

Such impressions have a basis in fact. It is largely because a number of civic-minded patrons, like the famous Medici, together with a handful of architects ruled by a modern vision of space were striving for just such an effect that the town now offers this lovely unified picture. Early in the fifteenth century the architect Brunelleschi was asked to top the cathedral with the cupola that now dominates city and hillsides, gathering all the loose ends of rooftops and spires into the gleaming apex of its orange-tiled shape. (The construction of the cathedral, a matter of great civic pride, including the mighty cupola, had been achieved by committees of experts representing the whole range of Florentine

Brunelleschi's sun-lit interior of the Church of San Lorenzo, Florence, begun about 1419, completed in 1446, after Brunelleschi's death — one of the earliest examples of the new architectural style.

skills, at times even by popular referendum.) Brunelleschi was also commissioned to build several smaller churches on both sides of the Arno, as if to balance the weight of the cathedral; these enhanced the feeling of a harmonious whole. Patronized by progressive businessmen, most of them bankers, sometimes by guilds or the city-republic itself, pioneer builders like Brunelleschi, Michelozzo, Alberti were dotting the city with structures proclaiming the new sense of space through their houses of worship, private chapels, residential *palazzi*, or charming villas scattered across the surrounding countryside.

The Church of San Lorenzo, just behind the cathedral, is the earliest full-fledged example of the new style. Built by Brunelleschi on a commission from Cosimo Medici and a few other wealthy businessmen, its interior was designed to capture the sunshine, to infuse the religious devotion with a serene sense of worldly well-being. The sixteenth century completed this ambitious conversion of a typical Medieval town into an early model of a modern metropolis.

Yet oddly, as one strolls down from the hills into the streets of the city, that sense of organic perfection is gradually lost. Even more strangely, one seems to be stepping from the lucid heights of the Renaissance back into Medieval confusion. The houses reveal their simple design for the most part as Medieval, though here or there the pattern is interrupted by an occasional Renaissance palace or church. The streets are narrow, often sunless and crowded, exuding the musty smell of the ages, spiced by the exhalations of crafts that have been here for many centuries. The bustle, the babel of voices echoed by the ancient housefronts, the whole tenor of life is unmistakably Medieval. In fact, very little has changed here since the thirteenth century, amongst the sturdy façades of the buildings, often enshrining a tiny statue of the Madonna in some quiet niche, except that some parts of the old city were razed, and birth and death have done their inevitable work. Generations have come and vanished, changing the composition of the crowds without really affecting their life style or subduing their buoyant vitality.

The truth is that this most perfect of Renaissance cities reveals itself as an essentially ageless being, living and breathing continuously since Medieval days, a human and structural entity of almost a thousand years, for which the Renaissance supplied little more than a perfect setting. Few things express as palpably as a great city the perennial quality of human life when seen in historical focus — the way the patterns of streets, buildings, people, even their gestures and speech, outlive the limited number of years that is our individual lot. If at times we may despair at the feebleness of our powers pitted against our brief years and the turbulence of our time, the life of the cities appears to spell the unexpected comfort that some ultimate human substance lives on through the ages, each day reasserting itself with irrepressible zest. What is so abstractly referred to as "historical continuity" is, in fact, a remarkably tangible presence.

The life of the city of Florence was neither disrupted nor especially disturbed in its flow by Renaissance culture. It was merely heightened in a dramatic setting, raised up on a festive, widely visible stage. Yet the lifeblood of the new drama continued to be supplied by the people of Florence, who had been there since early Medieval times and, with their earthy and colorful ways, their rugged individualism and profound pride, have outlived the Renaissance to this day.

* * *

What is true of one Renaissance city is true of the Renaissance. The first impression is of a brilliant contrast with the world of the Middle Ages. On closer inspection one discovers how much there is of continuity. The later Medieval centuries were blending into the Renaissance. The Renaissance was the organic peak in a continuously flowing evolution, yet it marked one of the most drastic breaks in the historical continuum — a puzzling ambiguity for the historians. The Renaissance was a fierce anti-Medieval revolt fought with characteristic Medieval weapons. It was also the logical consummation of practically every major Medieval development.

Nowhere is this ambiguity more striking than in science. Nowhere does the Renaissance contrast more sharply with the Medieval phase, yet the boundaries are nowhere harder to define. Renaissance science contrasted with the preceding centuries in that it was moving on an empirical, perfectly lucid and rational track, past all the mysticism or murkiness of the Middle Ages into a brightly lit modern era. This was the first phase of modern science, exuding the same air of factual clarity and exactitude that has come to distinguish modern science throughout. Small wonder that the great intellectual explosion that ushered in the modern world, the Scientific Revolution, received its first impulses from the Renaissance and that several of its pioneers had experienced the immediate inspiration of Renaissance Italy. (Copernicus studied at the universities of Rome, Padua, Bologna, and Ferrara. Galileo was raised in and around Florence and spent his final years as an exile on one of the Florentine hills.)

On the other hand, the pronounced empiricism of Renaissance science was not something totally new. It was a logical sequence to the empiricism of Albertus Magnus, to the methodological awareness of Robert Grosseteste and his student Roger Bacon, to the technological experiments of the Medieval craftsmen or even those of the alchemists.* Basically, it was the strong Arab bent for the empirical detail that reached its fullest flowering in the Renaissance, whether in botany, geography, geology, pharmacology, optics, or anything else. Whenever Medieval science, as the immediate recipient of the Arab influence, came closest to the earthy detail, it was approaching the Renaissance. In the same way, the rational strand of Medieval science, especially the bold advances in fourteenth-century physical and mathematical theory, intensely influenced Renaissance thought.

For all that, Renaissance science cannot be understood on any purely sober, clinically rational level: it was impelled, in one great unprecedented powerful thrust, by amazingly emotional, unintellectual forces. The chief impulse behind Renaissance science was a kind of intoxication, an infatuation with nature and its detail. Renaissance scientists, as a rule amateurs, in the literal sense of the word, were in love with their subject — or, since the subject was nature, were driven by a passion to study every single one of its aspects. Botticelli or Leonardo studying the particular features of a rare flower or plant clearly felt all the intellectual excitement of the botanist, as well as the esthetic raptures of the artist. Their portrayals of flowers give classical expression to both: they are fanatically precise to the minutest detail and yet expressive of a great artist's empathetic reverence for nature's handiwork.

Art and science converged so closely during the Renaissance that they often became almost interchangeable. Many a time the modern viewer is unable to tell whether a given drawing should be seen more as an art work or a scientific study. Nor are we always certain whether to "classify" some individual genius as scientist or artist. Many a great Renaissance artist and architect

---

* Robert Grosseteste, circa 1168–1253, firmly established the study of science at the University of Oxford, where he taught in the early part of the thirteenth century, and of which he was chancellor (magister scholarum) in 1214. Grosseteste's initiative was all the bolder in light of the fact that the teaching of science, linked to Aristotle's name, was the subject of several explicit prohibitions at the University of Paris during the same period. Inspired by ancient and Arabic science, Grosseteste was a pioneer in comprehending and setting forth the fundamental principles of the modern scientific method, emphasizing the concepts of the inductive and experimental approach. (See A. C. Crombie, *Medieval and Early Modern Science*, volume 2, pages 10 and following, as well as Crombie's above-mentioned *Robert Grosseteste and the Origins of Modern Science*, Oxford, 1953.)

excelled in the sciences of his time, to which he made outstanding contributions. Leonardo da Vinci's is only the most spectacular case in a long line of Renaissance scientist-artists.

Why is it that feeling and intellect, these two types of mental activity that we like to distinguish so categorically, were so closely linked in the Renaissance? The reason lay in the historical character of the new culture — that is, in the process by which the Renaissance had come about and which gave it its essence and meaning. To put it simply — including many different aspects in a single glance — the Renaissance was a revolt of the entire human personality against everything in the Medieval tradition that tended to stifle elementary human needs. The revolt had its economic aspects (which gave it a solid footing in practical life) in the early capitalist rebellion against feudal restrictions. It was as long in the making as it was sweeping, affecting every facet of human activity for a period of several centuries. Thinking was just as important in this reassertion of the repressed human potential as the stirring of feelings. However, at the core of the revolt was the reassertion of nature as a vital aspect of human life, because Medieval repression had above all prevented the human personality from relating freely to the natural world. Opening their souls once again to nature's long-avoided touch, Renaissance people were as eager to use their minds as their emotions. In fact, it was "the world," including the natural world, that represented the all-absorbing experience of the Renaissance; analytical thought, precise observation, or esthetic delight were really no more than different means through which people's reunion with nature was achieved. It was a great festival of the "senses" — a favorite word that Renaissance people liked to use in its dual meaning.

At the height of the Renaissance, the Venetian Giorgione painted a scene in which the innermost strivings of the Renaissance seem to have been fulfilled. Some young men and women are enjoying themselves in the midst of a lovely Italian summer landscape of meadows and trees. The women are naked. One of the young men is expressing his pleasure by playing a lute, and the other is accompanying him on an instrument hidden from sight. One of the women, turning her beautiful torso to us, is playing the flute. While playing, they look at each other as musicians will, enhancing the musical harmony. Indeed, the whole painting exudes a feeling of harmony — no longer the harmony between heaven and earth of the Chartres cathedral, but between men and women, body and spirit, and, most of all, the human world and the natural world.

One has to be in an extremely sour mood not to feel like joining in this quietly jubilant festivity. A long, tortured struggle had come to an end with people's happy return to nature, from which they had been kept for so long. Science during the flourishing of the Renaissance became a major factor in a cultural revolution whose primary manifestation was art. Renaissance science was the intellectual aspect of the great human revolt against Medieval repression. In that sense it was innovative, relishing experiments, allied to the artistic instinct, proudly modern, sensitive, and bold. Yet in another sense and under a larger perspective, science had played that same role at least since the days of Chartres. The Renaissance merely climaxed a long-standing Medieval development, in the sciences as much as in other fields. Since the later Middle Ages, science had become an integral part of the revolution against the restrictions of tradition and history.

* * *

The "return to nature," the new vision of man and the world, which the Renaissance defined in its art, had a long history. Even traditional Medieval painters had indulged in occasional glimpses of the natural world, notwithstanding their otherworldly conventions. A medium that deals in the perceptions of the eye can perhaps never completely close itself to the temptations of visual reality, no matter how world-denying the cultural context.

Yet from the end of the thirteenth century, more than a hundred years before the Renaissance achieved its ultimate triumph, painting took a deliberate turn toward the human and natural scene. Rocks, trees, architectural details, cityscapes, seascapes, boats, horsemen, characteristic faces, were increasingly crowded into the frescoes and panels. Medieval sculptors had struggled with the challenges of the human body for an even longer span of time, with perfectly lifelike statues emerging before the middle of the thirteenth century, and plant and animal shapes invading the cathedral décor even earlier. Several generations of explicit anatomical and landscape studies, including the first clumsy experiments in perspective, prepared the ground for the brilliant mastery of nature in the art of the mature Renaissance.

It is clear that the awakening from the Medieval dream of the other world did not occur with a sudden start. The eye had to accustom itself slowly to the earthly surroundings. The sense perceptions adjusted gradually to the visible details of this world.

As one looks at the Renaissance at its height, all these careful preparations may seem no more than a matter of perfecting certain technical skills subsidiary to the creative purpose. Like a child learning to draw from life, the artists seem to have been absorbed in the task of capturing the true shape of plants, animals, houses, and human beings so that fully developed Renaissance art might range more freely over the earthly domain. In fact, what went on inside the studios from the late thirteenth century on, far from simply a case of the artists "learning from nature," was a vital facet of the historic change in the outlook on the world. It was the visual aspect of the process of cultural reorientation itself.

True, a great deal of teaching and learning was going on inside the studios. All through the Renaissance, artists were systematically developing their capacity for detailed observation to make up for the many centuries of visual neglect. Drawing from nature became an intensive part of an artist's apprenticeship, studying the detailed shapes of leaves, flowers, animals, rocks, and, above all and again and again, the human body. If anything, these studies — botanical, anatomical, geological, and others — became increasingly systematic, suggesting that art was focusing more and more closely on individual detail. From Giotto and the Sienese painters to Michelangelo and Leonardo the exact study of nature was a passionate part of the Renaissance.

But that was only one aspect. Through their studies Renaissance artists were actual pioneers in the cultural process of developing a new vision. For once in thousands of years, art was in the front line of history. It was in the studios, or during the artists' outdoor excursions to sketch in street or countryside, that the West was relearning the habit of using its eyes. Emerging from the darkness of a culture that, for all its glowing metaphysical visions, had neglected the tangible world, it was the artists who first got used to the daylight of leisurely observation and who taught the new habit to their contemporaries through their paintings. Broadly speaking, modern science could hardly have taken a single step unless this thorough visual training had been performed by Renaissance art, which unlocked the natural world for the human eye.

There was still more to it. In the noonday heat of a narrow Florentine back street, inside the cool vault of an artists's workshop, the passionate study of the natural detail became also an important phase in early modern intellectual history. It was due to this aspect, no doubt, that the leading minds of the age, the Renaissance humanists, cheered what they called "the new art," *la nuova arte.* To a degree difficult to imagine now, the new art sparked an intellectual excitement among the Italian public far beyond its esthetic scope. The intellectual meaning involved major implications for science.

There is telling evidence that the rise of Renaissance art involved intellectual developments not confined to visual problems. Although earlier art had neglected the representation of nature (so that wherever such glimpses appear, they often have something touchingly childlike and naïve about them), the painters had attained great mastery in their treatment of traditional subjects — the majestic figure of Christ as the World's Ruler, the heavenly figures of the Virgin and Child with saints. There is little that is primitive or naïve about the great treasures of Medieval art, whether mosaic or manuscript illumination, fresco, bas-relief, or stained-glass window. In fact, we are often amazed at the artist's skill in his treatment of details — hands, feet, gestures, facial features — let alone the overall expression and mood. Medieval paintings often reveal a delicate, almost abstract esoteric treatment fitting for their unearthly subjects, but they are not lacking in representational skill.

Hence, though it is true that the artists had yet to learn how to depict areas of experience that had long been ignored, the reasons for the neglect, as well as for the shift toward natural subjects, were really a matter of taste or, rather, of cultural preference. Both the perception and the representational skill were fully developed wherever the focus of the general interest happened to dwell. By carrying sample on sample to their studios to be drawn "from nature" — or by hunting with their sketchbooks for any unusual aspect of the daily scene — Renaissance artists were not just realizing the new vision; they were the avant-garde for a new attitude toward the world, even for a new philosophy asserting the importance of this life, such as the humanists were formulating in their essays and pamphlets. The eye, the "king of the senses," as Leonardo liked to say, and its perceptions were acting as pilot for the modern mind.

The humanists welcomed the new art with enthusiasm because they saw in it a fraternal movement paralleling their own efforts for a new world-affirming attitude. Petrarch, who spoke of Giotto with profound reverence and listed one of his paintings as a prized possession in his last will, had himself introduced into early modern poetry such earthy subjects as the beauty of a little river flowing swiftly between its banks or the moods of a beloved and flirtatious woman.* Boccaccio, with his lusty descriptions of urban life, praised Giotto, the avowed initiator of the new art, for his incomparable realism and closeness to nature, "insomuch that men's visual sense is found to have been oftentimes deceived . . . taking that for real which was but depictured."

These paeans continued throughout the 1300s. "This Giotto . . . brought [the art of painting] to the modern style; and he possessed more perfect art than any one else ever had," wrote Cennino Cennini in a popular handbook toward the end of the century. The writers of the high Renaissance continued the alliance with *la nuova arte*. Advocates of an enthusiastic affirmation of the tangible world, they recognized that art was infinitely more effective than the written word in singing the world's praises.

*   *   *

But was art really packed with all that historic meaning? Art's descent to earth might strike us as simply a charming game, played for purely esthetic reasons. When a fresco by Giotto or one of Simone Martini's enchanting panels draws us into its secret little world, all historical considerations tend to fall by the wayside before its sheer quiet magic. The lovely artistry, the disarming sincerity of the feeling, the innocence of the experimentation in scenic or anatomical detail, may make us forget that these quiet

---

* Both themes were to become favorite subjects in Renaissance paintings, the little river as a background feature.

works embodied a revolutionary concept. Yet what must have struck Giotto's public before anything else was that the traditional flat backdrop, whether silver or gold or just gloomy black, seemed to have opened up like a curtain rising before an elaborate stage set, revealing a view of sky, hillsides, and olive trees, of Tuscan countryside or crowded Florentine streets.

Physical space as the public knew it from its daily experience had been admitted into the two-dimensional painting by some startling device. The foreground might show certain individual objects — a vase, a half-opened book, even some architectural structure — in their natural depth, to heighten the observer's feeling of looking into actual space. Voluminous figures might fill the foreground, diminishing in size toward the rear. The viewer, in other words, was witnessing a conquest of natural space and all that it contained, or at any rate a visual, conceptual conquest, a method of capturing space on a two-dimensional plane — a feat as remarkable for its scientific implications as for its sheer art.

Oddly, the method used by the artists to convey this illusion was not the application of the laws of perspective, which were not formulated till the 1430s, about a hundred and thirty years after these space-conquering efforts began. Nor was it really any particular technical gimmick at all. Rather, it was a conceptual development, the reflection of a new attitude of the mind. Traditional Medieval paintings suggest the surprising fact that the earlier Middle Ages had lacked a fully developed concept of physical space. Although this seems hard for us to believe — let alone visualize — the evidence of both art and science appears to confirm it. For hundreds of years art had presented the sacred figures as if they were without actual weight or substance but were floating in an ethereal kind of vacuum. This was quite in line with the accepted scientific theories, which did not allow for physical volume within the heavenly sphere. It was the physical aspect of the transcendental view of the world that perceived earthly matter as heavy, "material," and basically unclean, and the objects of the heavenly beyond, by contrast, as pure, esoteric, and sublime — like the quintessence, which according to Aristotle made up the essence of the stars.*

* See footnote on page 185.

At roughly the time the new art began to appear, physical science entered on a revolution. In a series of crucial new starts, theoretical thinkers, physicists like Jean Buridan and Nicole Oresme, of the so-called School of Paris, and the mathematician Thomas Bradwardine, at Oxford, began to re-examine certain fundamental elements of the system that had so far been taken for granted — raising questions about the nature of motion, impetus, gravity, and so on. The upshot was that fourteenth-century science was beginning to see the heavenly spheres as subject to the same physical laws as the "sphere" of the earth, with the result that the whole universe could be viewed as a physical entity — while the earth itself, by the same token, revealed itself more and more as ruled by universally valid physical laws. Concepts like motion, volume, or space, which had a severely limited meaning under the Aristotelian system of physics, acquired their modern significance only through this critical process. The new physical thought freed them from their stunted and crowded condition for a life of universal validity. It was the beginning of a development — without doubt the most momentous phase in the entire history of science — that culminated in the Scientific Revolution, when a solid terrestrial globe (indeed a "sphere" in the literal sense) was found to revolve around the sun, together with the other planets and according to the same universally valid laws.* Early modern physics was replacing the transcendental cosmos, including its intrinsic physical duality and essential lack of substances, with the homogeneous universe we have inhabited ever since.

Whatever influence these incipient scientific concepts may have exercised on art — or whatever common cultural developments may have been behind both — fourteenth-century painters, beginning with Giotto, most emphatically treated the setting of their sacred stories as empirical, physical space, not as the ethereal substance that had formed the backdrop for art until them. In all likelihood, by the fourteenth century people had come to think of space in the sensuous, tangible way they knew from their daily experience, at the same time extending the concept to the universe. The new feeling, which was at first probably unconscious, was expressed in art and only gradually led to the complex new

* See footnote on page 131.

approaches in science.* Artists all through the fourteenth century were eager to use every conceivable device to evoke a sense of natural space — including certain types of foreshortening that, though still far from accurate, helped to convey a three-dimensional impression.

Giotto especially displayed an inexhaustible ingenuity in finding ways to suggest physical space surrounding his sacred figures. A series of papier-mâché-like rocks thrusts toward the rear, as a backdrop from his *Flight into Egypt*. A smiling young woman's face peaks out from under a slanting little doorway as she receives a bundle of swaddling clothes from a friendly neighbor in his *Birth of Mary*. St. Francis' brethren crowd around his deathbed, pressing and pushing each other in their overwhelming grief.

* Giotto's first major series of paintings, his frescoes in the Arena Chapel in Padua, was completed around 1305. The critical review of Aristotelian physics was begun by mathematicians at Oxford's Merton College in 1328, and continued by the theoretical physicists of the School of Paris later in the fourteenth century. Reasonably enough, the new feeling, really a new cultural attitude, preceded the intricacies of a detailed scientific critique.

Giotto's FLIGHT INTO EGYPT, from the Arena Chapel, Padua (circa 1305).

Detail from Giotto's BIRTH OF MARY, from the same cycle.

Architectural details — a city gate, a domed church, the interior of a room — might vaguely recall his hometown of Florence, sufficiently so as to place the sacred tales in a familiar everyday setting. Some hundred years later Masaccio went further and let his St. Peter walk through an unmistakable Florentine street, with its rusticated housefronts and wooden galleries much like those we still find in the old parts of Florence.

In a purely esthetic sense all this must have been a thrilling experience, aside from the conceptual implications. Our senses have been dulled by the habit of six hundred years to the delights of the first loving depictions of physical volume and space — a turret astride a rooftop, a receding housefront broken by a little balcony or a bay window, the exciting contrast between foliage and masonry. Contemporaries like Boccaccio were particularly impressed that it was all so "completely lifelike," as if one could walk around and touch whatever was shown. A kind of sensuous satisfaction must have emanated from these intimate revelations of ordinary reality, coupled with a blissful feeling of coming home.

But what about the public? As some new fresco was unveiled in a church; as word got around and the people flocked there come Sunday — were they rooted to the spot, mouth gaping, hands folded in motionless awe? For a few instants, maybe. But meanwhile, in their minds, they were no doubt moving along on this journey of exploration, from detail to detail, from corner to corner, from one town to the next. Suddenly, the familiar appeared as though seen through the traveler's impressionable vision. For the public, the artists were cleansing the daily surroundings of layers of centuries-old visual neglect, revealing even the most trivial in the light of fresh discovery.

The fact is, the exploration of space by the Renaissance painters involved the viewer in a vicarious form of travel, an effortless way of moving about the earth, a mental facsimile of physical motion. It was not only the actual voyages of discovery, with their cargo of exotic impressions, that represented an obvious organic extension of this experience — so that the domestic landscape paintings, together with the maps and illustrated travel reports of the Renaissance, formed a kind of panoramic tableau of the world, unfolding before the age's insatiable visual curiosity. The same mood reflected itself in a different medium as well. The appearance of motion in Renaissance sculpture and bas-relief, following in due course the painters' discovery of physical space, depicted people casually standing, strolling, galloping on horseback, or immersed in milling or fighting crowds — a dramatic change from the essentially immobile figures of the Gothic statuary and all the long centuries of static Medieval art.

The world, near or far, had been unlocked by the painters. Now people could feel they were moving about the world by their own efforts, simply by identifying with the new avant-garde sculptures: Donatello's *St. Mark*, representing the proud old workman of the drapers' and woolworkers' guild, casually letting his weight rest on one leg while relaxing the other, a man pausing for a moment at his workbench or on his way home from a working day. Or Ghiberti's *St. Matthew*, who represented the guild of the bankers with an even more sovereign casualness — the Renais-

Donatello's ST. MARK (1411–13), representing the drapers' and woolworkers' guild, proudly stands in an outside niche of the church of Orsanmichele, Florence.

sance banker, proud of his ringing successes, on his way to or from his banking business.* Or, if a merchant should have occasion to travel to the city of Padua, he might find another one of Donatello's sculptures, the *Gattamelata,* and identify with a warrior on horseback, a symbol of dynamic physical strength. In purely psychological terms the discovery of the third dimension achieved in Renaissance art must have brought with it an incredible sense of liberation.

*   *   *

But the breakthrough also had its objective significance. By achieving a major step forward in art, painters had implicitly performed an important service for science. They had created a medium through which the various empirical sciences could henceforth demonstrate their findings with unprecedented visual plasticity. To a degree never attained by ancient or Islamic art, and certainly never even approached by the traditional art of the Middle Ages, painting and drawing could from now on be utilized for the illustration of scientific texts or as an effective device for recording minute observations.

The representation of objects in their true volume, for which early Renaissance art opened the door, lent itself superbly for the demonstration — and at times even the investigation — of empirical detail. Once art had taken the step from the flat (or essentially substance-less) to the three-dimensional vision, any object or figure could, in principle, be reproduced in a complete, all-around manner, with its accurate volume and proper proportions, so that every detail appeared in its legitimate place. By using several paintings, the viewer could make the object "revolve," so to speak, on the canvas or sheet of paper, and by turning it this way or that, observe more closely the individual features. The three-dimensional reproduction, in a sense, proved to possess an identity of its own, which was sometimes more revealing (as well as being easier to manipulate and more permanently preserved) than the

* Both figures belong to a series of statues protruding from the façade of Orsanmichele, an unpretentious church like a stunted skyscraper in the heart of Florence. The Florentine guilds decorated it with figures representing their various crafts, so that any passing guild member would naturally identify with the self-confident posture of the statues in their niches above, or the lively action of the bas-reliefs underneath the statues.

Ghiberti's ST. MATTHEW (1419–20), representing the powerful CAMBIO or bankers' guild, occupies a niche in the same façade.

actual object. It had acquired the properties of a scientific "model." Just as a cut-out cannot be shown from all sides, the flat conception of the figures in a traditional Medieval painting would have precluded these flexible uses. Renaissance art had, in effect, unlocked the third dimension for the scientific imagination. The emergence of scientific illustrations were merely a practical use of this conceptional step.

Some Renaissance artists were particularly aware of the possibilities of the new medium they found in their hands. With meticulous effort and considerable personal sacrifice, Leonardo developed it into a major tool for anatomical demonstration. Anatomical studies were a consuming passion for him. In his *Notebooks* he speaks of the "fear of living in the night hours in the company of those corpses, quartered and flayed and horrible to see," which he, the fastidious esthete, had nevertheless determined to dissect with painstaking care.

"I wish to work miracles," he writes in the same context, going on to explain his scrupulous dissecting procedures. But he was referring not just to the general value of anatomical studies for art. Supreme scientist and artist that he was, he had fully grasped the scientific significance of the three-dimensional drawing, recognizing its value for putting into focus details direct observation was likely to miss — much in the manner of a modern slow-motion camera or a microscope. "And you who say that it would be better to watch an anatomist at work than to see these drawings," he wrote in his notes, "you would be correct *if it were possible to observe all the things which are demonstrated in such drawings in a single body in which you, with all your cleverness, will not see or obtain knowledge of more than just a few veins . . .*"

In fact, his drawings and comments, when collected in one massive volume, present a complete course of anatomical study. They may easily be divided into systematic sections ("Osteological System," "Myological System," "Genitourinary System," and so on), which may again be subdivided (myology of trunk, of head and neck, of shoulder region).* All the drawings, with their infi-

---

* This has now been done in a magnificent volume (*Leonardo da Vinci on the Human Body*, edited by Charles D. O'Malley and J. B. de C. M. Saunders, New York, 1952), demonstrating the systematic scope of his anatomical studies. Leonardo himself had for years collected drawings and notes with a view to pub-

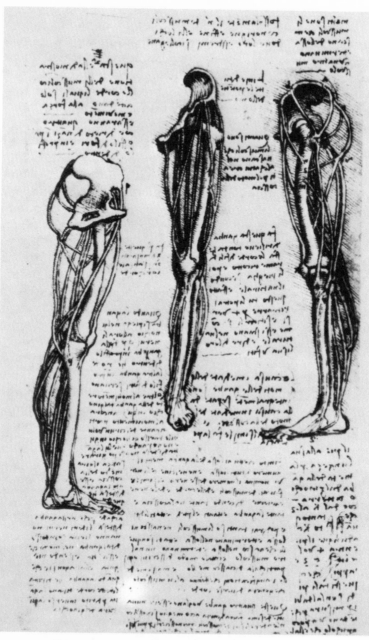

Leonardo's studies of the anatomy of a left leg are among the illustrations for his anatomical notes.

nite, patient detail, bear the imprint of his powerful, masterly stroke: one of the most artistic hands in history was working in the service of a great scientist's passion for the detail and his ambition to achieve total completeness. Moreover, at every step he is quite explicit about using the three-dimensional method ("Let the lungs together with all the spiritual members be shown from four aspects").

Toward the end of the Renaissance this development of a branch of art into science was climaxed by Andreas Vesalius, who made dramatic use of the new medium in a series of famous drawings. Although Leonardo's direct influence on Vesalius has not been fully established, the Flemish physician — or his illustrator — was clearly building upon the cumulative knowledge of anatomy amassed by Renaissance art, to which Leonardo had made by far the richest, most systematic contribution. In the illustrations for Vesalius' physiology, a sequence of skeletons is dramatically set against a somber mountain scene (the locale is the Euganean Hills, on the Italian mainland behind Venice). Dangling from their gallows, tossed to and fro by the wind, the macabre corpses exhibit the complete structure of tendons, joints, and bones, which Leonardo and his forerunners had uncovered and recorded. One should note that the historic value of Vesalius' work is now primarily seen in these illustrations, as an explicit visual introduction to the study of anatomy, for which his text supplied the comments.*

---

lishing a definitive treatise on the subject but was never able to complete that project — an index of the central tragedy of his life and work. About 1513 he wrote what he intended as an introduction, containing this characteristic passage: "... Here will be shown to you in fifteen entire figures the cosmography of the Microcosmos in the same order as was adopted before me by Ptolemy in his Cosmography. Likewise I shall then divide each member as he divided the whole into provinces..." (Ibid., page 32). The passage seems to imply that Ptolemy's *Geography*, almost one hundred years after its translation into Latin, still served as a model for scientific treatises during the Renaissance and that geography ("cosmography") was considered a kind of mother science to the emerging new sciences of the early modern age.

* Vesalius' *De humani corporis fabrica* (usually referred to as the *De Fabrica*) was published in 1543, the same year as Copernicus' *Revolutions of the Heavenly Spheres*. It is interesting to note how far both these pioneering works of modern science were influenced by the Italian Renaissance — Copernicus'

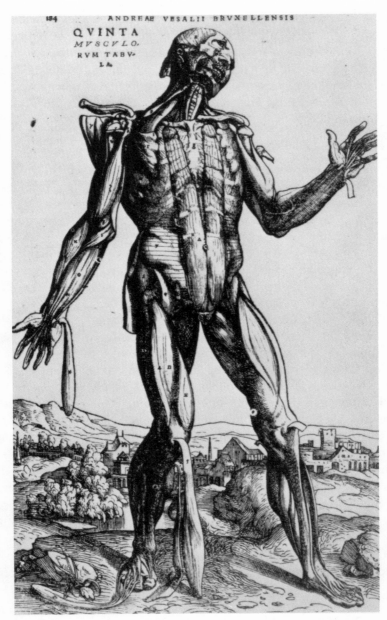

Illustrations from Andreas Vesalius' DE HUMANI CORPORIS FABRICA, 1543.

The breakthrough into the third dimension, spearheaded by Giotto and expanded by the conscientious studies of many subsequent Renaissance artists, whether painters or sculptors, and climaxed by Leonardo da Vinci, had produced the foundations of the modern science of physiology. The use of Renaissance art for scientific demonstration attained its greatest triumph in the study of the human body.

*   *   *

Virtually all empirical sciences profited immensely from the new dimension in art. It gave them a medium, both for explicit demonstration and for the storing of observations, that in its significance may be compared to the role of mathematics in the theoretical disciplines. Not geography, geology, mineralogy, nor zoology, botany, pharmacology, or physiology (to name only some) could possibly have developed to their modern level without the graphic, in-depth representation of samples based on the Renaissance concept of physical space.

The new medium played a particularly important role in the rise of modern geography. Besides the plastic representation of individual features on relief maps (of which quite a few were produced in an elementary form toward the end of the fifteenth century), the rise of the modern map itself, with its capacity for projecting curved segments of the earth in their correct proportions upon a piece of parchment or paper, was decisively furthered by the skill of Renaissance painters in handling problems of space through the use of perspective.

Initially, the influences had worked the other way around, with science leading the way for the artists. Leon Battista Alberti, the superb Florentine architect who formulated the laws of perspective for painters around 1435, had in turn been inspired by the cartographers' craft — or, more precisely, the rules for map projection he found in Ptolemy's *Geography*. In explaining various types of projection, Ptolemy had touched on the problem of fore-

---

through his encounter with the new cosmological theories that led from Toscanelli to a number of astronomical pioneers Copernicus had met personally during the ten years he studied in Italy; Vesalius' through his debt to Italian Renaissance art. (The unknown illustrator of the *De Fabrica* is thought to have belonged to Titian's studio.)

---

shortening and suggested a method for its cartographic representation. Alberti realized the underlying problem for artists was basically the same. He took a clue from Ptolemy and applied his mathematical concept to the type of foreshortening involved in the simple visual perception that is represented in a painting. Generations of Renaissance painters and bas-relief sculptors from then on could indulge in the accurate depicting of depth, the problem with which artists had wrestled since the time of Giotto, introducing a major dimension of realism and drama — all thanks to a cartographic concept.

Map projection and perspective painting, in other words, were developing along parallel lines in a mutually fruitful relationship for some hundred and fifty years, or across most of the Renaissance.* In fact, the method for projecting the curved surface of the earth on a map while keeping actual distances in their proper proportion, and the painter's technique in capturing perspective images on a canvas in the exact geometrical order in which they appear to the eye are two ways of dealing with the same problem. Whether one wishes to represent a large portion of the earth, or merely the tiny sector the eye comprehends at a glance, the mind is, in principle, faced with the identical challenge. In both cases a window is opened, so to speak, on a segment of earthly space, whether small or large — just as Giotto had "opened a window" on the world when he broke through those bland backgrounds of the traditional Medieval paintings — and the three-dimensional image presenting itself to us is then reduced to flat shapes on a two-dimensional surface, in a way that suggests the true distances and proportions. (Alberti in his book on perspective actually suggests that the problem facing the painter may be reduced to the appearance of three-dimensional objects on a windowpane.)

The problem in both cases was to come to terms with the three-dimensional nature of physical space, a problem that had been characteristically alien to the traditional Medieval mind, with its essentially abstract way of thinking about physical reality. Offhand, the mapmaker seems to be free of the particular geometrical

* Ptolemy's *Geography* had appeared in Latin translation in 1410, twenty-five years before Alberti first published his laws of perspective, while the basic modern map-projection techniques were worked out in their final form by the Flemish cartographer Gerard Mercator around the middle of the sixteenth century.

problem besetting the artist — that is, the apparent converging of lines in the observer's eye. But though this particular "subjective" factor is missing when one is constructing a map, the necessity of reducing curved shapes to a flat surface involves at least comparable distortions. It is significant that, except for an occasional circular outline, maps prior to the Renaissance had ignored the curved nature of the earth's surface. It was only during the fifteenth century that mapmakers began to suggest more explicitly the spherical shape of the earth on their world maps, and only during the sixteenth that they learned to cope with the distortion of distances resulting from the curvature of the earth when plotted on a flat sheet.

In a very real sense geography developed a fully three-dimensional perception of the earth only during the Renaissance.* Both the cartographer's projection and the painter's perspective were extensions of Giotto's breakthrough into physical space, refinements of his pioneering achievement.

* * *

There is a room in the Uffizi Gallery in Florence containing a painting by Giotto on one wall, facing a painting by Cimabue, Giotto's teacher, on the other. Both paintings deal with the same subject, executed along almost identical lines — the Virgin and Child on a throne, surrounded by angels. Yet between the two paintings extends the gulf of an artistic revolution. Where Cimabue's sacred figures are residing in heaven, Giotto's Madonna has come down to earth. Cimabue achieved his ethereal effect by placing his figures in the traditional Byzantine manner, flat against the gold background, so that they appear to be suspended in midair, towering above the viewer. He also elevated the Virgin's

* It is worth noting that Ptolemy's remarks about alternative methods of projection were not sufficient to lead to the immediate emergence of the modern map. What promoted the evolution of mapmaking from Medieval to modern was without doubt the new "plastic" vision developed during the Renaissance. Maps — along with the new geographic consciousness typical of the age of discoveries — developed parallel with the evolution of landscape painting, indicating a common underlying awareness of the world. The link is illustrated not only by the expanding sense of panoramic vision progressively expressed in Renaissance landscape painting but, most strikingly, in Leonardo's "jump" from his background visions (as in his *St. Anne* or *Mona Lisa*) to a conceptualized series of maps (discussed below), which seem to originate in the same panoramic sense of landscape.

Left: Cimabue's VIRGIN AND CHILD (before 1285), Florence, Uffizi Gallery.
Right: Giotto's VIRGIN AND CHILD (circa 1310), Florence, Uffizi Gallery.

throne by showing glimpses of heavenly space underneath its base and elongated the picture through various devices, for good measure. There can be no doubt that in this altar panel we are meant to look upward at the Virgin on her throne, floating high above earth.

The Virgin has lost none of her divine grandeur in Giotto's panel, but the grandeur is in her expression and bearing, not in the way she has been placed. Her throne rests solidly on the ground. Its base looks sturdier, thanks to its elaborate marble decoration. The whole throne is presented as an object of substance and depth, built like a Gothic niche of the type the viewer was likely to see on any day in the streets of Florence. Where the older painter made his figures seem flat, pressed flush with each other against the shallow background, the pleasing rhythm of their postures the only visible movement, Giotto shows real people with individual expressions and three-dimensional bodies separated by palpable space, crowding in spontaneous postures around the throne all the way toward the rear.

Other paintings by Giotto may illustrate more dramatically his use of the third dimension, but the juxtaposition of these two treatments of the same, perfectly conventional, subject by master and student demonstrates better than any single work the enormous step taken by art within one generation. To see the Madonna brought thus down to earth must have seemed shocking to a great many people. Precisely because Giotto was using his innovative approach for a hallowed theme, his visual revolution must have hit them with the force of a jolt.

The Renaissance rooms of the Uffizi have been arranged in historic order so that the whole course of Italian Renaissance art seems to unfold before one's eyes as one strolls through the gallery. After the great leap from Cimabue to Giotto we see art leaving behind the world of spiritual symbolism and gradually becoming a delightful mirror of the world of secular reality, even in the still-copious religious paintings, which often enough look like conventional pretexts for introducing a new artistic experiment or a mundane subject. In room after room, painting after painting, we watch the freshly discovered dimension being explored through residential interiors and country scenes, with hills, rows of cypress trees, sparkling rivers and winding roads, elegant furniture and sumptuous clothes. In the end, our entire earthly habitat seems to have been encompassed by Renaissance art.

The "discovery of the earth" amounted in a very real sense to an exploration of the third dimension. It was here that science came to coincide most closely with art. The artists had to cultivate not only the realistic representation of natural detail and the study of perspective, but also that of anatomy and movement. How people moved around in their world, making themselves at home, was certainly a significant aspect of the new three-dimensional emphasis, subjectively perhaps the most important, in that it provided for the viewer an especially pleasurable type of self-identification; and anatomy had to do with the three-dimensionality of the human body. In areas like these the scientific investiga-

Opposite: Sketches of the motion of water are repeatedly interspersed among Leonardo's notes (written in mirror writing).

tion of nature and the artist's (and viewer's) esthetic delight often came so close that they were virtually overlapping. In Leonardo's numerous sketches of the motion of water it seems impossible to say what fascinated him more: the problem of defining the patterns behind that motion or its infinite esthetic appeal. The borderline between artist and scientist had become exceedingly fluid.

Renaissance artists loved to indulge in urban-renewal schemes — obviously much needed to break out of the gloom of the Medieval city. It was a veritable fad, and it produced not only a whole spate of city-planning pamphlets, city descriptions, and city maps; but also a series of painted cityscapes, usually conceived at such a revealing angle that they form an evident anticipation of the modern pictorial map. Nevertheless, they were paintings — sometimes used for decorative purposes — representing a definite stepping-stone in the evolution of modern cartography.

As a matter of fact, painters were often attracted to the map-making trade. The rapidly spreading new science clearly appealed to their stimulated visual sensibilities, and a few minor artists embraced it as a welcome source of additional revenue. Leonardo, probably for his personal pleasure, sketched several colored maps of sections of the Italian countryside that look as if they were executed as mere conceptualizations, without any help from the customary cartographer's tools. As almost always was the case, Leonardo was not alone in playing this ingenious conceptual prank; he was merely bolder than all the others. Panoramic segments of landscape appear often in Renaissance paintings, in a way that is reminiscent of a huge relief map — conveying the impression that the Renaissance eye perceived both smaller and larger segments of the terrestrial scene on a similar scale. If one compares an early landscape painting, like Ambrogio Lorenzetti's vivid country scene in his *Good Government*, with an early relief map, the dividing line between painting and map appears surprisingly thin.

It was no doubt from a similar sense of kinship between the details observed close to home and the attractive world out there that the imagination of artists responded to the esthetic stimuli of the age of discoveries. Renaissance art came to reflect the growing geographic consciousness through some of its favorite themes — depicting exotic countries and their people, or wallowing in the

Ambrogio Lorenzetti's GOOD GOVERNMENT IN CITY AND COUNTRY (1338–39), one of the frescoes in Siena's Palazzo Pubblico, offers a delightful opening into the busy life of a fourteenth century town and the surrounding countryside. Beyond displaying a new, proud consciousness of one's native urban scene, the fresco, especially in its rural part, suggests a panoramic vision that approaches the concept of a pictorial map.

thrill of travel itself. From the time of Giotto art loved to show faraway places — luscious palm trees and desert sands; strange buildings with colorful façades, reflecting the influence of Islamic decoration, dark-skinned people, often with startlingly authentic negroid physiognomies; exotic animals, like dromedaries, camels, and monkeys.

The age of the great explorations was ushered in by a general excitement about foreign cultures and distant lands. The contact with Islam had undoubtedly given the first impulse to this collective travel fever that seems to have infected the whole population, high or low. From the later Middle Ages on, it produced a highly popular brand of travel adventure books, travel fantasies, and actual travel reports, like Marco Polo's famous *Travels*. By the time the Renaissance was accepted as the dominant culture — first in Italy, later in all of western Europe — this popular current became fused with the direct reports of the discoverers, which aroused the

Under the often used guise of retelling the story of the magi attending the birth of Christ, Gentile da Fabriano's ADORATION OF THE MAGI (1423), Uffizi Gallery, Florence, shows a multitude of travelers from distant parts of the world arriving on horseback and by other contemporary means of transportation, avid to catch a glimpse of the sacred event.

imagination with their incredible facts. Meanwhile, the popular feeling, everybody's vague urge to exchange familiar surroundings for the exotic world far away, survived as a theme in Renaissance art.

* * *

In one of the first rooms of the Uffizi, in accordance with the date of its creation, 1423, hangs a fascinating panel, the *Adoration of the Magi*, done by a young painter from Padua by the name of Gentile da Fabriano.* What it really shows, under the pretext

---

* The panel was originally commissioned by the banker Palla Strozzi for a chapel in the Florentine Church of Santa Trinita.

of the pious theme, is the adventure of foreign travel. The exotic elements are all there — the monkeys, the people with dark skin and outlandish features, even a leopard and a lion. But before any of these details, what strikes the viewer is the great, powerful, joyous movement of the long train of travelers, crowding all the way from the distant horizon to the foreground. Clearly it is the act of traveling itself that the artist intended to celebrate, the unlocking of unknown places by the strength of one's own vigorous motion, whether on horseback or camelback or by boat. (A vessel can, in fact, be seen near the horizon.) Painted as the age of discoveries went into high gear — Prince Henry's Portuguese captains were in the midst of their systematic forays down the African coast — Gentile's panel echoes the feelings common among the people of the Renaissance that inspired the great expeditions.

The subject became a favorite in Italian art. Botticelli painted several such *Adorations* toward the end of the fifteenth century; Leonardo himself painted another. Always the overriding theme is the joyous excitement one experiences in foreign travel, personified by the rush of a traveling crowd approaching the foreground. To be sure, the time for popular travel was still some five hundred years away. Many profound changes had yet to occur in Western society before foreign travel became accessible to ordinary people. But Renaissance art reflects one of the chief historic motivations behind modern mass travel — the powerful longings of a people who had long been frustrated with the sedentary ways of the Middle Ages. Like a prisoner's craving for the outside world, the long-pent-up urge to break out of the confining orbit of the Medieval town seeped into Renaissance art, motivated the great discoveries, and provided the impetus for geographic studies, one of the earliest of the modern sciences in its own right.

*   *   *

The close affinity between art and science was not only due to the discovery of the earth, the fascination with a world that had in effect been forbidden; it also had to do with one of the most central features of the Renaissance, *l'uomo universale*. The artists' easy switching from an esthetic to a scientific approach and back was part of the phenomenal versatility that the Renaissance released through its appeal to the creative potential. After all, what could be more stimulating for the unfolding of one's talents than a cultural climate that called for uncovering the beauty and the hidden mysteries of this earth? A person's whole potential found itself pitted against the adventure of the world. Everything — art, science, technical skills, the mastery of different media, and a broad range of intellectual abilities — were called into action by this total challenge to the creative personality. Switching back and forth between different media became as common as the frequent overlapping between science and art.

All through the Renaissance, the gifted were proving to themselves — and to others — that they were still more gifted than anyone suspected, not only in their chosen fields but in other professions as well. Giotto the painter designed, and for a time supervised, the building of the lovely bell tower of the cathedral, the campanile, which is now one of the three widely visible landmarks of Florence, together with Brunelleschi's orange-tiled cupola and the fierce watchtower of the Palazzo Vecchio. Today, anyone looking down from the hills notices first the slender, gleaming structure created by the first painter of the Renaissance, without giving much thought to this remarkable switch between the media. (If one looks more closely, one notices that the painter showed his hand by using the multicolored marble of the campanile in a particularly playful, decorative way, so that the structure looks almost like one of those architectural forms Giotto liked to scatter across his paintings.)

Brunelleschi, greatest among the architects of the Renaissance, loved to show off his versatility in defiance of the conventions of the guild or others. Though trained as a goldsmith, he accepted the prestigious and technically challenging job of crowning the roof of the cathedral with a dome. We hear that he thereby aroused the resentment of the masons' guild, which had the trespasser thrown in jail. Freed by the city fathers, who were more appreciative of his freewheeling talents, Brunelleschi not only performed a major technical feat by bending the ribs of the Gothic vaults into the shape of a dome similar to the Roman Pantheon's, but, reverting to his earlier trade, he also tried his hand in a competition for the doors of the Baptistry, across the way from the cathedral. Yet when his design for a series of bas-reliefs was selected together with one by the goldsmith Lorenzo Ghiberti,

Brunelleschi refused to collaborate with his colleague on the project. By now, the city fathers, irked by the difficult genius, decided to assign the Baptistry exclusively to Ghiberti and, moreover, appointed Ghiberti cosupervisor for the cathedral dome. Finally, after what one would imagine was a great deal of angry quarreling, the two geniuses decided that each should stick to his proper job. Ghiberti did his so well that he broke through the confines of his own medium by developing his bas-reliefs into a series of masterly "sculptured paintings," done in a pioneering new style.

Brunelleschi's engineering achievement, topping the efforts of citizens' committees that had tried to solve the problem for many years, amounted to an important extension of the Gothic structure, with its use of statics inherent in ribs and vaults. But where the Gothic builders had used the static power of the ribs to uphold a relatively small vault made up of intersecting pointed arches,

Ghiberti's sculptured self-portrait peeks out next to his famous bas-reliefs at the east door of the Baptistry in Florence — a vivid testimony to the satisfactions of the creative artist.

Brunelleschi employed the same force to hold an enormous cupola in place. By a stroke of genius the basic principle of Gothic engineering had been extended into the first major architectural feat of the Renaissance. Yet what is remarkable is that, in looking toward the cathedral from the surrounding hills, one thinks as little about engineering problems as about the fact that the campanile was designed by a painter. Both structures in their simple loveliness appeal to our immediate feeling. The shift from one medium to another, or from engineering problem to sublime work of art, was achieved so naturally that it seems as if it had not taken place at all. To the versatile mind of the Renaissance artist there existed no barriers rigidly separating one discipline from the rest.

* * *

Such versatility appears to have expanded with the progress of the Renaissance. Even Michelangelo, who may seem an unlikely example, with his stubborn pride in being the world's greatest sculptor, could display surprising flexibility. For one thing, he performed a number of architectural feats: a nobly designed façade amidst a cluster of residential *palazzi*; a fountain in the center of a cloister garden, remarkable for its mixture of sturdiness and grace; a style-setting staircase, foreshadowing the Baroque; and, after all, the dome of St. Peter's, dominating the Roman *campagna* far into the Alban Mountains to the south, much as Brunelleschi's cupola reigns over the environs of Florence.

Like Leonardo, Michelangelo devoted himself to anatomical studies. The first modern art historian, Giorgio Vasari, who knew Michelangelo in his later years, tells us how the young artist received permission from the prior of Santo Spirito, one of the churches Brunelleschi had built, to use a sacristy for dissecting "many dead bodies." Like Brunelleschi or Leonardo, Michelangelo possessed a keen interest, coupled with ingenuity, in technical problems. Vasari relates how he managed to transport the towering statue *David* (the Florentines called him *"il gigante"*) from his temporary workshop in the cathedral to his destination in front of the City Hall. Together with a couple of friendly artists, Michelangelo constructed a huge wooden frame from which "the giant" was suspended by a slipknot and rolled to his destination through the streets of Florence, with the help of windlasses and

Michelangelo's DAVID now stands in a dramatic site in the Academy Museum in Florence, but was originally designed for the square in front of the Florentine City Hall, the Piazza Signoria (a copy is still standing there). The sculptor transported the enormous statute from his studio to its original site with the help of an ingenious contraption.

beams. Later, when he worked in the Sistine Chapel, he designed an ingenious new type of scaffold, avoiding the customary method of drilling holes in the ceiling and suspending the scaffold by ropes.

The famous ceiling frescoes of the Sistine Chapel are a final testimony to his versatility. A familiar anecdote shows the characteristic ambivalence of his personality — stolidly proud of his uniqueness and yet capable of amazing dramatic change.* The episode — probably true — is of Michelangelo's indignant refusal of Pope Julius II's request to paint the enormous ceiling of the Sistine chapel — *"ma io sono scultore!"* — until at last he locked himself in the chapel, picked up a brush, and began covering the ceiling and upper part of the walls with his brooding, gigantic figures. Working there for the next four years, he even went through a substantial change in his style, passing from the forceful brushstroke the sculptor employs for his sketches to the calmer, more confident treatment of broad, spacious planes suited to a painting. If the greatest sculptor since ancient Greece was to paint, he would show the world how to do it! Ghiberti, whom he had admired as a boy, had bent the medium of the bas-relief until it acquired the qualities of a painting; Michelangelo was now stretching the meaning of painting until it encompassed the strong points of sculpture as well.

In our age of narrow specialization we have become embarrassed by the possession of many different talents, as if we were worried about what pigeonhole to fit in for society's sake. Renaissance people reveled in their God-given talents. Both art and science benefited immensely from universal man's happy versatility at that early stage.

* * *

Leonardo had an immediate forerunner in the fifteenth century. A modern historian has called him the "universal genius of the early Renaissance." Leon Battista Alberti worked in cartography

---

* The historians have been busy sorting the anecdotes from the facts, as they should. Meanwhile, we may keep in mind that a historical anecdote, even if it should turn out to be an invention, is usually a perfect reflection of how someone was thought of by his contemporaries. *"Se non é vero,"* the Italians say, *"é ben trovato!"* Even if it didn't exactly happen that way, it very easily might have.

Michelangelo's GOD CREATING THE SUN, from his ceiling frescoes in the Sistine Chapel, Vatican, Rome (1508–12), forcefully demonstrates the transformation of the great sculptor into a painter.

and mathematics, besides formulating the laws of perspective and exploring other principles of architecture and art. To each field he made important contributions. Oddly, the modern science of coding, cryptography, also goes back to him — a sign of his astonishing intellectual grasp.

Although Alberti's name has not become a household word, the man with the fine Tuscan features conveying his vibrant nervous energy fully epitomizes the range of the Renaissance mind. Modern wartime intelligence agents, bent on cracking an enemy code, are as beholden to him as are art students, who still draw according to Alberti's rules. Receptive tourists, strolling from the center of Florence toward the river down Via della Vigna Nuova, are suddenly stopped in their tracks by the elegant sight of the most perfectly proportioned façade in this city. The Rucellai Palace, perhaps the most harmonious building in Florence, is from his hand.

Can we with our timid little specializations still envisage such a vast reach of the mind? Surely, despite the evident nervous tension overlying his face, he must have been impelled by a vision of the ultimate harmony of all life. The truth is that the universal genius believed in the universal unity of the world. Alberti himself saw this in mathematical terms, as a principle of order inherent in the structure of the universe, the idea of beauty implicit in the natural order itself. In fact, it was probably a last glimmer of Medieval universality, seen through the rational mind of the Renaissance.*

Perspective, where Alberti made his most incisive contribution, was the product of almost a century and a half of collective experiments. Painters from Giotto to Masaccio and Paolo Uccello, sculptors like Ghiberti and Donatello, architects like Brunelleschi himself, at times assisted by Paolo Toscanelli, had engaged in all sorts of practical experiments and occasional theoretical studies before Alberti could formulate the exact laws in a mathematically valid way in his book *On Painting*, published in 1435. A major scientific breakthrough was prepared not by a handful of qualified scientists, but by a long line of practical artists whose social status was basically still that of craftsmen. Instead of proceeding from

* The Medieval concept of *ordo mundi*, integral to Chartres's natural philosophy and embodied in the Gothic cathedral, was carried over into Renaissance thought in many ways and on many levels. The difficulty was that the Medieval sense of the inner unity of the world was centered on an idea or, more exactly, the sphere of ideas that constituted the "other world" of Medieval Christianity and philosophy (as well as the invisible dimension of the mystics); while the Renaissance, having moved the central focus of its world-view from heaven to earth — and from the divine to the human level — was confronted with the inevitably fragmented world of empirical objects. The problem of how to envision a universal unity behind a world perceived in individual material terms is still with us, since empirical phenomena tend to evoke in us pluralist impressions, and unity can only be conceived in terms of ideas. The great Renaissance artists clearly strove to preserve (or rekindle) a feeling of universal harmony in their paintings — Botticelli through a kind of all-enveloping mystical atmosphere, probably inspired by Neo-Platonic thought; Michelangelo in the Sistine Chapel, by re-creating the Christian universe in individualized terms that were acceptable to the Renaissance; Leonardo da Vinci, perhaps disturbed by his unending involvement in empirical observations, by conjuring up a concrete universal vision in his paintings. (Marsilio Ficino attempted to build a system of philosophy, based on Neo-Platonic thought, designed to encompass the universe in the natural self-realization of the human mind and will.)

Leon Battista Alberti's RUCELLAI PALACE in Florence was the first consistent attempt to apply classical orders to a Renaissance palace front.

the theoretical study of optics to the specific problems with which the painter contends, the artists themselves were groping on canvas and bas-relief, or with the help of primitive models, till they hit upon the solution.

The laws of perspective, in other words, were arrived at in a process of trial and error, of exchanging experiences among work-shops; and the "scientific" formulation occurred as a kind of after-thought, an abstract summation of the practical experiments. Since pretty much the identical development — deriving a general principle from a vast amount of experiences amassed through practical work — took place in certain parallel fields, we may say that Renaissance science was distinguished by the first major en-counter of technical practice and scientific theory. The two main strands in the evolution toward modern science, which had met only very casually and infrequently in the past, became at last entwined in a relationship that was often to modify its nature, but remained inseparable from this point on.

A parallel development was taking place in geographic studies. Toscanelli and his friends, after all, developed their new concept of the earth from the practical experiences of Portuguese seamen, sifting these empirical data through the theories of Strabo and Ptolemy, as well as their own keenly logical thought.

Again, there was a striking parallel in the twin inventions of printing and engraving, where a lively exchange of ideas and results among German workshops preceded the actual "invention." In-terestingly, all these strategic advances seem to have reached their final acceleration during the 1430s or early 1440s. It was a water-shed between Medieval and early modern science, a crucial time when the practical legacy of the Medieval workshop (and of Medi-eval seamanship) was fused with the theoretical tradition of Medieval science, with its scholastic, mystical, and classical strands, and modern science became firmly welded to its empirical base.

* * *

In establishing the laws of perspective, Alberti was acting as a typical Renaissance scientist: with a perfectly trained mathemati-cal mind he sifted the groping advances of the artists, which were sometimes brilliant but on the whole unsystematic and halting. Like Toscanelli, he probably visited them in their studios, located in their narrow Florentine streets, usually opening toward the greenery overgrowing a backyard. Being a fellow artist or architect as well as a trained mathematician, he undoubtedly questioned and argued and debated as one of their own. He grasped the essential meaning of their experiments as well as their remaining unresolved problems, related them to the legacy of classical optical theory from antiquity through Islam, with which he was familiar, and reduced them to a lucid and readily applicable set of concepts.*

Alberti's achievement is best illustrated by the development of the "focal point" or "vanishing point," which plays such a decisive role in perspective. Earlier Renaissance painters used to let the lines converge in several points rather than a single focus. (In one painting by Taddeo Gaddi, a follower of Giotto's, we notice such a plethora of focal points that the intricate structure serving as a backdrop looks more like a maze at an amusement park than a portion of the old part of Florence.)

By Alberti's time a sophisticated artist had come to assume that there were two focal points to a view, thus introducing at least some degree of order into the original chaos. Brunelleschi is credited with the theory of a single, unified vanishing point. In fact, about eight years before Alberti's *On Painting* was published, a mural in the Church of Santa Maria Novella, the *Trinity*, had been unveiled, in which Masaccio stunned the Florentine public with a perspective view of a chapel that seemed so perfectly lifelike that for an instant the viewers thought they were looking at the real thing.

By sheer groping, perhaps with a little occasional expert help from a mathematician like Toscanelli (who seems to have been intensely involved in these experiments), the Florentine artists had come as close to solving the ancient problem of depicting perspective as was conceivable in a purely pragmatic way. What Alberti did was to take the decisive step beyond this inspired experimentation. He established that the vanishing point is a function of the human vision and that its geometrical site may be obtained by imagining a triangle whose base coincides with the artist's eye and whose sides converge toward the triangle's point. (Alberti saw the triangle as projected upon his imaginary veil, or grid.)

* For instance, he introduced an imaginary veil, or, as we may say, "grid," between the eye and the object, upon which the image projected its angles of perspective and on which they could easily be traced. (He might have gotten that idea from the "grid" of longitudes and latitudes, with which the geographers around Toscanelli seem to have begun operating around this time.)

A less technical explanation is that Alberti raised a long-standing tradition of haphazard practical experiments to a valid scientific level by translating it into the proper geometric terms, and thereby introducing it into the science of optics. He also brought definitive order into the highly chaotic kind of vision from which the artists appear to have suffered when they first opened their eyes to the world. Most remarkably, he created that order by frankly centering it on the perceiving individual, the cardinal point of the Renaissance.

For all its technical complexities, Alberti's seems a lovely achievement, worthy of someone who saw the world in terms of a mathematical order, and worthy of the man who designed the most harmonious building in that most harmonious of cities.

* * *

Leonardo's eyes in the famous self-portrait are the strangest eyes in the world. Apparently focused in intense scrutiny and yet looking into an infinite distance, they sit like a pair of jewels in a face lined by age and the furrows of bitter experience. One is tempted to call them the eyes of a visionary, except that his visions were always emphatically of this world. As a scientist, and of course as a painter, Leonardo believed in vision as the empiricist's primary tool. Yet of just what manner was that vision, just how the world looked through those extraordinary eyes, was an enigma to his contemporaries and is still a mystery for the scholar today. Though he seems to have left us a unique record of his vision of the world in his paintings and was presumably more articulate than any other artist in recording his thoughts, both his art and his ideas, far from solving the riddle, raise endless new questions, absorbing us deeper into the maze of one of the most unusual minds of all time.

For all the reams of learned studies that have been written about Leonardo, we still seem unable to say what he so restlessly sought in his observations with sketchbook and notes. How are we to understand a mind that united the scientist's analytical curiosity with the most refined sensitivity of an artist, both on the genius level and both pursued with such relentless absorption? The fact that his art and his scientific thought show a continuous interweaving — like his notes on the nature of light, or his exhaustive anatomical studies, or his copious geological observations, all of which are abundantly reflected in his art — merely compounds the puzzle. It suggests almost inescapably that there was some vital connection between the two in his mind. But if that was so, just what was that connection? Amazingly, the riddle of this mysterious man who at the same time seems so intensely human, so close to what we all strive for in our intellectual being that we believe to know the answer in our hearts, is unsolved; scholars continue to feel provoked by — and write provocatively about — the puzzle of Leonardo. Perhaps he will remain a puzzle, and as such one of the favorite subjects of interpretive scholarship of all time.

Leonardo da Vinci does not fit into any of our accepted categories. That could be because he possessed too much vitality himself — because he was too intensely absorbed in life, which he studied without pause. He was neither the coldly detached scientific observer, serenely assured in his superior analytical stance, nor the passionate artist mindlessly enamored of his subject. If anything, it is in his carefully, and indeed "scientifically," constructed paintings that he conveys a supreme detachment; it is in his scientific notes that he often speaks with the voice of the youthful enthusiast, forever in love with his theme, like the classic, perpetual amateur. And yet, though a serene sense of calm pervades his paintings, a kind of cosmic harmony into which the great nervous restlessness and the nameless problems intrinsic to his art seem to have been resolved, one can hardly call him cold. Nor can one think of Leonardo's scientific observations, despite their fervent, often poetic quality, as in any true sense "amateurish." They reveal a supremely original mind, on many telling issues incredibly advanced beyond his time, fertile with the most astounding insights and guided by a solid methodological feeling — though, again, one cannot say offhand precisely what that method was.

His refusal to fit into existing categories — which disturbed his contemporaries as much as it puzzles us — extends to almost any facet of his personality and life. In his ideas he remained outside the mainstream of scientific thought, a forever-amazing peculiarity in such a formidable mind. Not only did he lack the academic training that would have put him on an equal footing with other scientific thinkers of his time — so that, despite the avid reading by which he tried to overcome that handicap, he often missed the exact formulation of a problem (and sometimes the solution) that had already been achieved. What isolates him even more drastically is that he never made a single direct contribution to the systematic growth of scientific knowledge. (His influence on Vesalius' anatomy, if it was there, was in all likelihood indirect.)

All of Leonardo's countless observations remained in the stage of scribbled notes. Though he planned to write, and presumably publish, a major treatise on anatomy, of which at least an amazingly systematic framework has been found to exist, the sheer incessant rush of diverse observations appears to have washed over that ambitious project. What we have in his *Notebooks* is nothing but a rambling, rather erratic monologue, the conceptual shorthand of a titanic mind, relentlessly devoted to the contemplation of one phenomenon or problem after another, in a breathlessly alternating sequence as each claimed his attention. We must conclude that whatever absorbed him with such fervor must have been something he was studying for himself.

Part of Leonardo's mystery lies in the fact that this gigantic mind, which in a flash was able to anticipate some of the crucial insights of the Scientific Revolution — and a substantial segment of twentieth-century technology, to boot — belonged to a gigantic loner. To compound the puzzle, Leonardo wrote his notes in mirror writing. It was not that he had to hide his thoughts from anybody's secular police or from any ecclesiastical censorship. The power of the papacy did not extend into the proud Italian city-states like Florence or Milan, where he lived and worked. Neither the contemporary Church nor local rulers like Lodovico Sforza, the Duke of Milan, were in the least unsympathetic to his inquiry. The reason for Leonardo's secretiveness was peculiar to the workings of his mind: in jotting down his observations he was communicating with himself and did not want anyone to glance over his shoulder. His notes are the reflections of an extraordinarily intensive intellectual process that had to shun the public light while it was taking place.

* * *

It was as a loner—and a mystery—that Leonardo moved through his time. Even when a young man — strikingly handsome and of athletic strength — he became the target of suspicious gossip. Whether there was any basis to it or not (he was accused of homosexuality), his response seems touchy, considering how permissive that age was. Although the charges were dropped, Leonardo seems to have lost his taste for living in Florence and shortly after found employment with the Duke of Milan. Throughout his life, he was touchy, too, on a different score. His notes reveal him as forever asserting his intellectual competence against some invisible rival, someone no doubt in the assured possession of an academic degree. This famous man, then, with his dual genius as scientific thinker and artist, whom one might imagine as forever basking in the public's esteem, was plagued by a lifelong sense of social inadequacy, some inferior feeling toward the "normal" world, no doubt caused by the incongruent workings of his genius. A man whose personal fame, great in his lifetime, is almost unequaled in history seems to have suffered without let-up from a need for society's recognition for his sexual preferences (surprising in a Renaissance artist) and, even more amazing, for his qualifications as a scientist.

In actual fact, Leonardo's observations were undoubtedly a great deal freer and bolder precisely because he was unencumbered by all that Scholastic clutter of contentious dogmas handed down inside the universities. Though his private reading was extensive enough to inform him of the more important results of Medieval science, he kept the freshness and immediacy of observation, which a regular academic training in the Scholastic tradition almost certainly would have suppressed. One of the mainsprings of his scientific genius, an important element of the modern viewpoint he brought to science, was precisely that he was able to look at the world with little more than his own fresh, unbiased, incomparable vision.

The notorious "Leonardo mystery" is constantly deepened by his contradictions. His homosexual leanings, which became more obvious in later life, seem contradicted by his most famous painting, the *Mona Lisa*. Some art historians, bothered by the apparent inconsistency, have suggested that his model was really a man.

Leonardo's MONA LISA (circa 1503), Louvre Museum, Paris.

But the model, to the best of anybody's knowledge, was in truth the young wife of a considerably older Florentine businessman who had to spend a great deal of time out of town and, rather thoughtfully, one feels, provided his young wife with the distraction of having her portrait painted by the famous Leonardo. For all we know, the thoughtful husband, Messer del Giocondo, may have felt quite safe because of those rumors that were going around town.

But events do not seem to have worked out that way. The sittings in the lady's home took up a full three years, with a regularity that seems entirely out of keeping with Leonardo's usually erratic working habits. During this time he kept refusing (so we are told by people other than Leonardo) commissions from kings, princes of the Church, and noble ladies. Very likely the bored young wife, who, shortly before the sittings began, had suffered the trauma of a miscarriage, found her frequent sessions with the handsome and celebrated painter an effective distraction indeed. For lack of any other documentation, the famous portrait, with its notorious "enigmatic smile," distinctly shows us a woman in a silent erotic interplay with an unseen male partner, no doubt the artist himself. The mystery Leonardo has captured is the elusive subject of erotic tension playing between a woman and a man. Not only the subtle flesh tones of the lady, the fluid smile, the total vividness of someone who won't, or can't, sit still — an amazing element to capture on canvas — suggest the silent uproar of the senses; even the background conveys the same mood. Like a surrealist symbolization of sensuous passion, the wild and stormy background scenery in all its sweeping depth has been divided into two utterly unequal portions, with even the horizon broken up in two asymmetric halves; the view loses itself amidst the suggestive colors and shapes of an erotic dream landscape.

The famous enigmatic painting by the mysterious genius seems to reveal its secret once we abandon our preconceived notions and open ourselves to the true feeling it evokes. Leonardo has tried to penetrate one of the most evanescent mysteries we can experience — but may experience nevertheless on any day — and has succeeded in capturing it for eternity.

* * *

It is tempting to end our story on a mysterious note, tempting yet deeply unsatisfying. Leonardo's presumptive enigmatic figure lends itself easily to the kind of intellectual game scholars have long played with him and, in a sense, with the whole of Renaissance science. It is always safer to dramatize unsolved problems than to suggest answers. Yet under a larger historical perspective, much that seems bizarre about Leonardo appears really in a far softer light. Once we view him as a product of the Renaissance or, even more sweepingly, of a continuous four-hundred-year process, the solitary genius seems to lose much of his appearance as an enigma and a loner, history's perpetual maverick. Instead, Leonardo looks more like the perfect symbol of that evolution — mysterious, yes, as far as the inner workings of genius are bound to retain an ultimate element of mystery, but the logical sum total of all the preceding developments, nevertheless.

The historical context reveals, first of all, that Leonardo's dual mastery of science and art was not unique. That duality was profoundly endemic to the Renaissance, a twin manifestation of its emphatic concern with the world, in retaliation for the traditional otherworldliness of Medieval culture. Not only is the earlier Renaissance scene alive with instances of scientist-artists on a more limited scale; the frequent flexible ranging between different media itself was an integral part of the personal ideal of Renaissance culture, the *uomo universale.* Leonardo shared his keen pride in his versatile talents — "I wish to work miracles!" — with virtually every single creative figure of the Renaissance. The discovery of the human potential — which in practice meant the creators' discovery of their own selves — was an organic facet of Renaissance culture, a natural aspect of its central adventure, the discovery of the world. A passion for self-discovery and self-realization is among the most potent motives of the Renaissance. Even Leonardo's insecurities over his lack of an academic status may well have been a by-product of that collective passion. At least, whatever personal experiences may have initiated these feelings, they were certainly strengthend by the cultural trend.

All these are marginal features. The important thing is that Leonardo personifies the very essence of Renaissance culture, its fascination with life itself. If this central cultural challenge had released formidable energies before, it provoked a total response in Leonardo — total in that fascination with life completely absorbed him throughout his lifetime and was of a unique, indeed phenomenal, intensity.

* * *

It took a culture destined by its historical position to focus on the elementary experience of the world to produce someone so utterly absorbed in contemplating all its manifestations: How do rivers originate? What kind of geological structure do we observe from inside a cave? How do birds power their flight? How does a fish propel itself by its tail? How does the hare use its hind legs in running? How or by the use of what muscles does a man who is sitting on the ground raise himself to his feet? And so forth and so on, as life showered him with its questions.

Problems of motion held an incessant interest for Leonardo, just as they were in the center of theoretical science in his time, but so did cosmological problems, geological formations, questions of meteorology, anatomical details in animals and humans.

His mind, to judge from his *Notebooks*, seemed almost feverish in its restless activity. More likely, it functioned with an extraordinary balance and harmony, healthily alternating between concentration and rest, but in its active phases working with an absorptive force that remains astounding. Perception and creation went intimately hand in hand. So did his science and art: he sketched while jotting down observations, often to illustrate a point. Frequently his notes related directly to his painting, in that he desired a more minute understanding of a visual phenomenon — the way the sunlight's reflection on a leaf depends on the leaf's transparency; the way a shadow is cast over an object or human face; the structure of a particular bone. The observer of nature was helped by the painter; the artist by the scientist. Such a large portion of his notes is at least indirectly related to problems of painting that one may argue Leonardo pursued his studies for the primary aim of putting his art on a "scientific" foundation. (His quest for a more legitimate social status may have motivated him in this.)

Yet any such one-sided explanation ignores the phenomenon of a mind completely absorbed with its subject — the world in its totality, life itself — using every conceivable device for penetrating its secrets. Whether it was any one of the myriad specific riddles nature was posing for his observing mind or something more pervasive and fundamental — some quintessence or spirit of life, like the evanescent quality of erotic attraction he tried to capture in the *Mona Lisa* — he was clearly caught up in a perpetual search. At the heart of the mystery of Leonardo is a passionate search for the mystery of life itself.

* * *

If this is how Leonardo appears in the light of history, why, finally, do we think of him as the archetype of the modern scientific mind? Why modern? And why a scientist at all? Once again, the answers are far from self-evident; but they may net some final insights that could make them worthwhile.

On the face of it, Leonardo's status as a scientist seems doubtful. How can one be considered a scientist without making any tangible contributions to the advancement of science? (His contribution to the rise of modern anatomy and physiology, let us recall, was in all likelihood indirect rather than firsthand.) What is more, there is the problem — of which Leonardo himself was only too keenly aware — of his tenuous grip on the existing body of scientific learning, because of his extremely sketchy education. His decidedly outsider status certainly excludes him from that sturdy band of natural philosophers who, since the days of the School of Chartres, had furthered the early modern world's understanding of nature through a consistent and eager exchange. Even his astounding technical ideas cannot really be considered "inventions," in the absence of any practical manufacturing applicability. As we meet them in his secret sketches and notes, they strike us as incredibly brilliant flashes, prophecies of a phenomenal scope, but not as an organic part of the evolution of modern technology.

A similar problem tends to veil his modernity: Leonardo seems at first far more Medieval than modern. The prophetic vision of technological gadgets many centuries away — the motor car, the submarine, the airplane, the helicopter — he shared with certain Medieval minds, in particular Roger Bacon, who preceded Leonardo by more than two hundred years and projected the evolution of modern technology in several identical features, although not with the same detail.

Closer to the essence of his thought, was not Leonardo's search for some ultimate force behind nature's manifestations a typical Medieval quality? When he jots down in his *Notebooks*, "The motive power is the cause of all life," he seems to imply more than that his search was aimed at some underlying substance (or "cause") in back of all of life's phenomena, quite in the tradition of the Medieval alchemists. By convincing himself that the force that originates motion is in fact that underlying cause, he comes surprisingly close to one of Thomas Aquinas' key arguments in favor of a divine presence behind nature. From Aquinas to the alchemists — indeed, from a long line of Medieval thinkers far too numerous to list — Leonardo had inherited the premise that some invisible power looms in back of the natural world and manifests itself in its workings. Nor, for all his constant emphasis on the rule of "experience," does he hesitate occasionally to express such metaphysical ideas. "Nature is full of infinite causes that have never occurred in experience," he writes between fragments of rather more earthly thoughts. By his phrasing he implicitly attributes to nature some metaphysical role, transcending the limits of sober empirical observation.

Then in what sense do we have a right to call Leonardo modern? The answer is: probably in every sense that has any historical significance.

He was undoubtedly an heir to Medieval science and knew it, despite the gaps in his knowledge.* In a more sweeping sense he was the culmination of the continuous Medieval and Renaissance tradition of science, as well as the first brilliant example of the modern scientific mind. More than anyone else, Leonardo, with his universal approach to the world, embodies the evolution of scientific thought from its Medieval beginnings via the Renaissance to the present.

Modern, above all, was his vision. No trace of any of the earlier hesitation or confusion clouds his glance. Leonardo's eye, his

---

* His *Notebooks* contain frequent reminders for him to pick up this or that book on a particular scientific subject. As is to be expected, almost all this literature was by Medieval authors.

"king of the senses," is unfailingly clear, penetratingly analytical, always going to the root of everything he observes, whether rock formation, tree root, or waterfall, like the dissecting knife of a surgeon. But his vision is never coldly clinical or "reductionist," like a modern laboratory technician's; for all its minute precision it is essentially panoramic, binding all details together in one warm, sweeping, all-encompassing glance. One perceives the artist's delight in the infinite beauty of nature behind his panoramic landscapes and realizes that his penetrating observation of natural detail was at bottom inspired by the same excited, invariably enthusiastic love for the world. Both ways his love extends into infinity — down to the secrets of the tiniest particular and upward to the vast distant horizons.

His first known landscape, a drawing he did when he was twenty-one, showing the Arno Valley near where he was born, already exudes that feeling. He had climbed a minor mountain in the neighborhood so that the widest possible view would open before his eyes. It is a midsummer day, and the still, clean air, the patches of sunlight, the clumps of trees, the very vastness of the land extending silently underneath the summer sky, evidently thrill him. He has captured the lovely peace of a summer scene with a "master's graphic shorthand," as someone has said, a few sparse strokes.

Renaissance artists had frequently painted their landscapes in a panoramic setting, but Leonardo seems to compel his viewers to look ever more upward and outward, in a constantly expanding panoramic view. Going beyond the incongruous dream landscape of the *Mona Lisa*, or the infinite horizons he evoked in his *St. Anne with Virgin and Child*, Leonardo proceeded to his conceptualized maps, visualizing an entire area as encompassed in a single glance. In one further sweeping vision, contained in his *Notebooks*, he visualizes the earth as it would appear to someone looking down from the moon. More than that: in the same brief entry he establishes, some thirty years before Copernicus, a vision of the solar universe that is in essence that of our modern perception, coolly rejecting the Aristotelian concept of the cosmos that had ruled for almost two thousand years.

From the quiet summer day with its view of the Arno Valley, his home, to his far-off horizons expanding into infinity of his

Leonardo's first known landscape, a sketch of the Arno valley near his birth-place, dated August 5, 1473.

imagination, to a view of the earth as seen by a modern astronaut within a Copernican framework, Leonardo's vision spanned the full range of the modern mind. Inspired by its matchless sense of beauty and loving warmth, we may see Leonardo's as a pantheistic vision, a way of worshiping the world for the intrinsic loveliness of its manifold forms of life. Though it had been a long road from the majestic harmony of the Medieval view, the divine unity of the world had in a way been rebuilt. There is perhaps no deeper religious experience conceivable for the modern mind.

Epilogue: The Tree of Knowledge

And the Lord God commanded the man, saying, Of every tree of the garden thou mayest freely eat:

But of the tree of the knowledge of good and evil, thou shalt not eat of it: for in the day that thou eatest thereof thou shalt surely die.

Genesis 2:16–17

THERE IS NO ENDING to our story, at any rate not a simple or happy one. What started as a stimulating intellectual pastime played against a fairy-tale background — from Arabic beginnings, via Medieval magicians and monks, to a climax in Renaissance Italy — has assumed the terrifying proportions of a monster. The catastrophic possibilities we have come to associate with modern science and technology may never fulfill themselves, but they are certainly there, frightening us with their shadow and, in many ways, with an already present reality: the corruption of our natural habitat, the physical corrosion of our cultural heritage, the manipulation of human life, the expansion of military technology, the dangers of industrial nuclear accidents.

Science has retained, perhaps multiplied, the intellectual and esthetic fascination it held for the people of the Middle Ages and Renaissance. No doubt the achievements of science since the Renaissance stand among the most marvelous adventures of the human mind, comparable in scope — if not in profundity — with the height of Greek philosophy or the metaphysical insights of the great religions. Modern science's staggering successes in unraveling the mysteries of nature have been an inspiration for humanity, by showing what the mind can achieve. But at the same time, science's contents have grown to such dimensions that the challenge no longer works for everyone with an alert mind, including the artist, the philosopher, the theologian, or simply the person with an interest in new directions and thoughts, as it did in these early centuries. The educated have become painfully aware that their individual culture — either scientific or humanistic — is likely to be severely incomplete. The lovely ideal of *humanitas*, of a culture encompassing both aspects of the "knowable" world, the ideal of the School of Chartres, has been eclipsed by the fragmented world-view of an overspecialized age.

Not all that stands revealed by this eight-hundred-year perspective appears necessarily in a somber light. Western science's increasingly intimate linkage with technology — begun during the Renaissance, nearly inseparable in most areas since the eighteenth and nineteenth centuries — intensifies science's power and lends it its cutting edge. But it also means that scientific ideas can bring immeasurable improvements in rapid course to those most urgently in need of them — homemakers, the sick and the old, the poor and disadvantaged. Only political or social barriers seem to stand in the way of fulfilling science's beneficial capabilities on a staggering scale — bureaucratic obstacles, institutional habits, vested interests, self-centered indifference. The potential for good is there as much as the potential for destruction.

Science could still help to make this an incredibly happier world, even as it could destroy it. Potentially a force for either infinite good or unimaginable evil, the problem of how to control its effects has clearly grown apace with its power over our culture. But the feeling that science has gotten out of control has never been as strong as it is now.*

* * *

Looking at our age from the respectable distance of eight hundred years is like looking at one's home through the wrong end of a telescope. In a way, one's own time takes on the faraway quality of a historical period. Emphases shift; values one has taken for granted appear, suddenly, in a questionable light; the whole age assumes characteristics one had never suspected.

In the historical perspective, our age stands out glaringly as the first civilization in history to be centered on science. Though

---

* A posthumous collection of essays by the late distinguished philosopher and historian of science Jacob Bronowski addresses itself to the problem of how the modern world can reassert its control over science.

science played a significant role in virtually all former cultures, including the prehistoric, ours — "modern Western," to be sure, but with its dynamic tendency for global expansion — is the first culture in which science has taken the central place as fountainhead and ultimate standard of cultural values, instead of religion, the devotion to the commonwealth or sovereign, or, more than religious piety, the absorption in otherworldly contemplation. Our uniqueness appears to make our civilization that much more of a historic experiment, with decidedly uncertain results.

Yet how did science achieve this prominent place? How did it gain its power over our culture during the four hundred years since the Renaissance? It did so not by its own strength alone, but because it was, in effect, invited. At some point in modern history, the people of Western civilization have surrendered to science an ultimate power over their minds and lives, much as the people of developing countries today are eager to welcome science and technology within their borders. Like these countries, early modern nations — France, England, the United States, Russia, Prussia, Sweden, Italy, in time the entire Western world — opened their arms to science, partly for exceedingly good reasons, having to do with its intellectual challenges and expected benefits.* Yet in part these reasons were based on a popular myth, a colossal historical misunderstanding, which is all the more difficult to unravel today because we still believe in it — because it is an integral part of our cultural heritage and accepted convictions.

The evolution of science, so the unspoken assumption suggests, has gone on in a consistent way all through history. Logically, its steady upward march has reached its highest historic peak now, with the future opening a view of further towering summits. Since we instinctively tend to equate science with rational conduct founded on incontrovertible evidence, it must fol-

* During the seventeenth and eighteenth centuries, most western nations furthered science through private or governmental institutions that expressed the popular enthusiasm, while the universities maintained a long-lasting stubborn resistance. The first modern laboratories and observatories were set up by government-sponsored academies (like the Royal Society in England or the Academy of Sciences in France), or by private discussion groups meeting in a collector's cabinet or a drawing room. It was groups like these that first promoted the use of the new scientific instruments and the dissemination of scientific tracts.

low that in some telling ways we must be incomparably more clever than our ancestors, who were doomed to live in varying stages of "superstition" and "ignorance." Science in this view — one that is, as a rule, unconsciously accepted — appears as the highest conceivable form of intellectual activity. Modern civilizations have submitted to its effective rule — just as modern individuals usually continue to do on a day-to-day, problem-for-problem basis — because the authority of higher wisdom or deeper insights can be neither argued nor denied.

* * *

Again, although a substantial part of these assumptions is obviously correct, the broader implications are not. Manifestly, science represents one of the most advanced forms of mental activity (*the* most advanced in matters of establishing a factual context). What is more, scientific thought possesses an innate tendency to advance from one insight to the next, which gives it an appearance of continuous dynamic progress. Nor is there any doubt that our age has achieved a vast superiority over the scientific achievements of any earlier civilization. For reasons stemming more from our cultural history than from the nature of science itself, modern civilization has aided the growth of science to an utterly unprecedented extent, with the result that we are continually witnessing spectacular feats of scientific progress. What seems more natural than to think that our civilization represents the ultimate triumph of the human mind in history?

There are quite a few things that are wrong with this concept. For one, the idea of continuous, dizzying human progress emanating from science would seem to clash severely with those cataclysmic fears that have attached themselves to the evolution of science in our time. How can we reconcile the notion of science's innate superior intelligence — if not, indeed, wisdom — with its destructive potential, revealing itself all around? The insoluble contradiction may be a major element in the mood of confusion and bleak despair that is often noted by modern commentators.

For another, the whole idea rests on an abominable perception of history. The growth of science represents nothing like an unbroken, steadily ascending progression carrying humanity on its shoulders, as it were. It would be far more realistic to think of the history of science as a continuous, sometimes desperate struggle against the periodic disruptions that the general course of history tends to inflict on the pursuit of consistent study (as we have repeatedly seen in this discussion). The resulting, often drastic, ups and downs were, moreover, accentuated by the notorious inhospitality of certain cultures to the systematic study of nature. The kind of collective escapism from the realities of the world that pervaded the West after the fall of Rome — and intensely tinged the character of Medieval culture, especially during the earlier centuries — forced science to a virtual halt, or at least to an almost unprecedented low level. In fact, even that minimal measure of continuity with the preceding evolution that is indispensible for the pursuit of science often had to be won against overwhelming odds, as the recovery of ancient science from the Arabic texts dramatically illustrates. To think of the evolution of science as a straightforward continuum, or to associate science automatically with progress, involves a profound misunderstanding of the relationship between science and history — if not of the nature of the historic process itself.

*　*　*

Yet if the power of science over the modern world is not the logical result of an undisturbed upward evolution of science across the course of history — combined with science's intrinsically superior intellectual gifts — how then did it come about? The answer must lie in the historical character of modern Western civilization.

The modern world — modern Western culture with its built-in tendency to reach out across the contemporary world — originated in a fierce revolt against the pressures of traditional Medieval civilization. The course of modern history has been indelibly stamped by that revolt, which is, in a way, still going on and of which we are still a largely unconscious part. (Our deep-seated resentments against authority in any paternalistic form, against restrictive family ties, and, on the other hand, our emphatic struggles for individual self-assertion and self-realization are living testimony to the durable hold of the Medieval heritage and the abiding continuity of that historic revolt.) In this sense the Renaissance itself is still going on. Its ideals and values are still very much our own; and the last five hundred years have served to spread the impulses of the Renaissance among different social groups, through different parts of the globe, and, above all, across different levels of individual assertion. Philosophy, religion, political and social thought, all were affected. On every level the individual began to cope with the repressive features of the transcendental legacy of the Middle Ages. From Luther's rebellion against the Medieval Church to the natural-law philosophers of the Enlightenment, to the formulation of the Marxist theory and program to the twentieth century feminist movement, modern history has featured a dramatic sequence of individualist revolts, each of them asserting the importance of life on this earth against Medieval world denial. The evolution of science in the Middle Ages and Renaissance, the whole process we have here let pass before our eyes, was clearly a significant chapter in that rebellious struggle. Just as St. Augustine turned against Greek science as a part of that material world from whose collapse he wanted to save his contemporaries, the revolt against St. Augustine's spiritual legacy called for the emphatic reassertion of science.

At the same time, the resumption of scientific interests had important ideological connotations. It not only presented a challenge for a sweeping exercise of long-dormant intellectual faculties, particularly in the realm of the immediate observation; it also involved a vigorous assertion of the "sensible world" itself, the whole observable world of nature that had been so explicitly banished by St. Augustine. On an intellectual level, the revival of science anticipated the emotional and esthetic affirmation of nature (and of the perceiving senses) in the Renaissance.

Yet science represented an especially outspoken and bold way of asserting the significance of the natural world. To insist, as the masters of Chartres explicitly did, that the study of nature is a God-given human right, or that nature is structured in this or that particular way and follows these or those particular laws, is a far more explicit act of rebellion than merely to point out the beauty of a landscape or a human body or face. In the course of the antitranscendental revolt, the revival of science amounted to an eloquent ideological statement. Modern science has retained some of these ideological overtones.*

* * *

In the long-range perspective, modern science appears to owe its prominence and unique development in the modern world to a peculiar set of historical circumstances. Since these somewhat remote conditions are obviously not very widely known, it is understandable that modern people tend to attribute the hegemony of science over our culture mostly to science's innate superior wisdom. Yet even the quality that has most strongly contributed to science's triumphs in the modern world, its unique set of methods — or, simply, the modern "scientific method" — appears ultimately as a product of these historic conditions.

Modern science derives its cultural leadership — including its implicit claim to serve as a yardstick for our private lives, a panacea for our most personal problems — from the pioneering role it has played in the revolt against transcendental restrictions, which ushered in the modern age.

It is because science played such an avant-garde role during the liberation from the repressive features of the Medieval legacy that it became, in effect, the ideological matrix for the modern mind. As the primary and most articulate expression of that historic revolt it was science that first raised its voice against the major transcendental taboos ("magic," "authority," "faith") and in favor of such distinctive modern tenets as freedom of inquiry, the basic lawfulness of the workings of nature, and the belief that human

---

* The birth of modern science is often linked to the rise of capitalism; but, as with most early modern phenomena, the social energies released by the incipient capitalistic system explain little more than the presence of a general vibrant atmosphere open to experimentation and change, contrasting with the economic gloom of the earlier Middle Ages. Early capitalist industry, even in its primitive Medieval phase, undoubtedly stimulated technology; but theoretical science, as we saw, had distinctly different origins, related to the philosophical need for creating a counterweight to the transcendental tradition. Generally speaking, early capitalist prosperity was no doubt an important factor among the conditions encouraging the antitranscendental revolt, but the revolt itself seems an inevitable reaction of the elementary human instincts against a cultural tradition that had severely repressive effects on the life of the senses. In other words, the rise of capitalism may have had something to do with the timing of the revolt that led to the Renaissance (including the renaissance of science), but not with its deeper causes.

---

nature, too, is subject to natural laws. It was in the course of that centuries-long revolt that the spokesmen of science, like Albertus Magnus, aggressively stressed the superiority of rational thought — the same approach that was increasingly used for unraveling nature — over all other modes of reflection. The strict emphasis on a rational and empirical approach — as opposed to the prevailing intuitive or mystical approach — was to have a decisive impact on the formation of the modern scientific method, as well as on the general values around which the modern attitude was formed.*

The general pervasion of our culture by scientific (and pseudo-scientific) values has been observed and often sharply criticized before, on many levels and from many different angles. In a way, the grave concern over a science-oriented rationalism and its limitations for the human mind goes back to Thomas Aquinas. It had its distinguished itinerary across the history of modern Western culture, via Kant and Rousseau, to such diverse twentieth-century thinkers as Alfred North Whitehead and Carl Gustav Jung. Always, the thrust of these critics, what they had in common over a seven-hundred-year span, was an uneasy — if not necessarily fully conscious — awareness that modern scientific rationalism, arising out of those late Medieval conflicts, involved a particular method and way of thought that was uniquely suited for its own purpose, but dubious if not outright dangerous for any area outside the scope of science itself. In a moving passage, written around 1270, Thomas Aquinas cautioned his readers not to lift the veil from those ultimate mysteries that are destined to be concealed from the human mind.† He gave this solemn warning,

* Even such a keystone of modern political and social thought as the philosophy of natural law received its first impulses from the new scientific consciousness of the Age of Reason, with its emphatic belief that the same forces that determine the workings of nature are behind the conduct of human affairs and must therefore be susceptible to the identical type of inquiry. We have by no means abandoned these convictions but, as modern psychoanalysis and, in particular, Freud's personal example show, are still intensely engaged in applying the insights and methods of the modern natural sciences to the area of human behavior. Indeed, psychoanalysis may be viewed as an outstanding example of the fertile influence of science on the whole field of human behavior — as well as its limitations.

† In book I, chapter VIII of the *Summa contra gentiles* Aquinas writes, citing Hilary: ". . . Yet pry not into that secret, and meddle not in the mystery of the birth of the infinite, nor presume to grasp that which is the summit of understanding: but understand that there are things thou canst not grasp."

which echoed across the next seven hundred years, having demonstrated with trenchant logic why the exercise of pure logical reason is bound to run up against insurmountable barriers. The greatest rational thinker of the Middle Ages, in other words, privy to the most complete scientific knowledge of his time, was warning his own generations and the generations to come not to overestimate the power of rational thought, but to acknowledge the superior scope of mystic intuition and sheer faith as paths toward understanding.

* * *

It still seems difficult for us to envisage history as an extension, a heightened dimension of individual human life, bound to follow familiar psychological patterns multiplied by the experience of entire cultures. But what happened in the emergence of modern science could easily be envisioned in human terms. Like an adolescent rebelling against a parent, early modern civilization had formulated its beliefs in deliberate contrast to everything the parent had held dear. Modern intellectual and cultural attitudes had their origins in the European people's fierce self-assertion against the established values of the Medieval world. A passion for the concrete objects on this earth replaced the transcendental visions. Mystic intuition gave way to the rigors of consistent rational thought. Instead of turning their eyes to the limitless orbits of the divine universe, early modern Western people had decided to stick to the empirical evidence, the carefully limited piece of specific reality relating to the matter at hand (with a grudging exception for mathematical proof). What the step from a Medieval to a modern vision amounted to was a spectacular narrowing of the focus — a sharpening and, at the same time, a coarsening — in our perception of the world. But the microscopic view, the ultimate triumph of this self-limiting vision, which accomplishes miracles when we are scanning the condition of our organs or any other physical detail, is bound to fail us when it comes to the subtleties of human behavior. Why people act in one way or another; how we ourselves ought to act; what are our or other people's innermost needs — such questions are not answered by magnifying the features of some particular physical object. The answers must be felt by an intuitive approach that is as open-minded to the whole human being as it is to the whole context of life. The Middle Ages knew that the important truths reside in a vast, invisible dimension, rather than in the visible features of some isolated physical detail.*

*   *   *

The landscape has changed drastically since Leonardo, on his hill, overlooked the Arno Valley. Is Western science, a story with ebullient beginnings, ending in confusion and tragedy, like so many modern plots? Perhaps; but that is not the inescapable conclusion our evidence implies. A more balanced result might suggest itself when one recalls that our present problem, the problem of science in our time, concerns the human control over science, the question of how we may make science serve well-defined human ends, extracting ourselves — or the essentials of our culture — from science's often oppressive power and its alarming tendency to develop headlong under its own steam. That seems to be the problem underlying the environmental protest; it is the social dimension of the problem; and it is the dilemma to which the historical perspective appears to point.

The sometimes invidious comparisons between the Medieval and Renaissance beginnings and the contemporary status of science that have been made here may suggest a few final thoughts regarding this issue. It may help us to know that modern science was not begun in a spirit of power involving unlimited threats, but in a happy and wholesome spirit, as part of a search for new mental horizons and a more fulfilling existence. It may also help to recall that science acquired its present power largely because of ulterior developments having little to do with science's essence and aims — in fact caused mostly by a rather elementary kind of overreaction against traditional authority, magnified on a historic scale. To deprive science of the more intimidating aspects of its myth — the myth that its power flows from the unmatched superiority of its insights and methods in all circumstances of life

---

* An interesting development tending to confirm these general impressions is the trend toward "holistic" medicine which is increasingly coming to the fore as a reaction against the growing specialization and mechanization in the modern medical field.

— or, in other words, to divest science of the ubiquitous authority with which we have surrounded it, might have a liberating effect on our thinking.

To look on science not as an institution beyond human control but as a phenomenon made by people for their enjoyment and, often, intense delight; a phenomenon endowed with increasing powers by people for extremely human reasons, no matter how far back in our history; in short, to be able to think of science once again on a human scale — such may be the modest lesson our history suggests.

Bibliographical Notes

Bibliographical Notes

Of the various works ranging over the whole period discussed in this book — some of which are sometimes cited on the order of "classic" historical treatments — the only one I found substantially useful is A. C. Crombie, *Medieval and Early Modern Science* (2 volumes, New York: Doubleday Anchor, 1959). Although far from presenting a definitive overview (perhaps a great deal of more specialized research will be necessary before that can be done), Crombie does offer an eminently helpful introductory survey, rich in valuable detail and with an extensive bibliography.

I have refrained from trying to approximate anything like a complete bibliography of the vast and steadily expanding literature on Medieval and Renaissance science. Instead, I am merely listing those studies which have been most directly helpful to me in preparing this book, with only an occasional reference to an article or book that seemed of major importance on a given issue. Sources for quotations occurring in the text are listed at the end of each chapter's note.

Chapter One: The Idea of the Earth in Renaissance Florence (pages 1–41)

For the geographic substance, see my "Geography in Fifteenth Century Florence," in *Merchants and Scholars: Essays in the History of Exploration and Trade* (ed. John Parker, Minneapolis: University of Minnesota Press, 1965), on which this chapter is essen-

---

tially based. The article contains a fairly extensive survey of the historical literature on humanism and the geographic discoveries, including Toscanelli's. The information on Toscanelli's gnomon is based on Gustavo Uzielli, *La vita ed i tempi di Paolo del Pozzo Toscanelli: richerche e studi*, pp. 284–285; 374ff.; and the concise summary, p. 600. A document representing a workman's bill, which appears to refer to the closing off of the lantern windows to facilitate the reading of the sundial, is reprinted on p. 600. For the construction of the lantern, see Peter Murray, *The Architecture of the Italian Renaissance* (New York: Schocken Books, 1963–70), pp. 30–31. A highly informative, lively glimpse into life in Florence during the Renaissance is offered by Gene A. Brucker, *Renaissance Florence* (New York–London–Sydney–Toronto, 1969). The fullest recent treatment of fifteenth-century Florentine art is Frederick Hartt's monumental *History of Italian Renaissance Art* (Englewood Cliffs, New Jersey, 1969?), including an extensive specialized bibliography. The role of the Medici in Florentine culture is extensively treated by Ferdinand Schevill, *The Medici* (New York: Harper Torchbooks, 1960, originally published 1949), a good overall sketch, even though outdated in details.

The process of cultural awakening to one's earthly surroundings is perceptively discussed by Kenneth Clark, *Landscape into Art* (Harmondsworth: Penguin/Pelican, 1949–56), especially chapter 1, "The Landscape of Symbols." Changing geographic concepts are impressively presented in their historical context by Norman J. W. Thrower, *Maps and Man: An Examination of Cartography in Relation to Culture and Civilization* (Englewood Cliffs, New Jersey, 1972). See also Lloyd A. Brown, *The Story of Maps* (Boston, 1950). A fascinating collection of oceanic and mariners' legends is W. Frahm, *Das Meer und die Seefahrt in der altfranzösischen Literatur* (Göttingen, 1914). The relationship between Renaissance painting, in particular cityscapes, and cartography is briefly discussed in my article "Impulses of Italian Renaissance Culture Behind the Age of Discoveries," in *First Images of America: The Impact of the New World on the Old* (ed. Fredi Chiappelli, Berkeley: University of California Press, 1976), volume I, pp. 29–30. Brucker's *Renaissance Florence*, chapter 2, "The Economy," includes a good picture of Florentine international trade, based on the existing wealth of specialized studies. I have focused on a few further details of Plethon's geographic discussions in Florence (including the map of the North) in "Some Reflections on the Origin of the Vinland Map," *Proceedings of the Vinland Map Conference* (ed. Wilcomb E. Washburn, Chicago: University of Chicago Press, 1971), pp. 50ff.

Strabo's conceptual statements are quoted from *The Geography of Strabo* (trans. H. L. Jones, London–New York: Loeb Classical Library, 1917), volume I, pp. 241, 243 (1.4.6). Polo's statement about the indivisibility of the "Eastern" Ocean occurs in chapter 6 of his *Travels of Marco Polo* (trans. Ronald Latham, Harmondsworth: Penguin, 1958–72), p. 248.

Dante speaks of the *"mondo sanza gente"* in the *Inferno*, Canto XXVI, 116. The text of the Toscanelli letter is quoted from *The Journal of Christopher Columbus* (trans. C. R. Markham, Works Issued by the Hakluyt Society, volume 86, London, 1893), pp. 4–5.

Chapter Two: Ancient Roots (pages 42–64)

Of the plethora of general histories of science flooding the market during the 1950s and 1960s, I found William Cecil Dampier's *A Shorter History of Science* (New York: Meridian, 1957, originally published 1944) the most unpretentiously useful. A stimulating philosophical approach (to an as-yet-wide-open subject) is R. G. Collingwood, *The Idea of Nature* (New York: Galaxy, 1960, originally published 1945).

A vivid idea of the transition from prehistoric to early historic life in the Near East is conveyed by Henri Frankfort, *The Birth of Civilization in the Near East* (New York: Doubleday Anchor, 1957), chapter 2. See also H. and H. A. Frankfort, J. A. Wilson, T. Jacobsen, and W. A. Irwin, *The Intellectual Adventure of Ancient Man* (Chicago, 1943, Penguin edition under the title *Before Philosophy*.) For a balanced assessment of certain controversial aspects see Sabatino Moscati, *The Face of the Ancient Orient* (New York: Doubleday Anchor, 1962), pp. 329–30. Benjamin Farrington, *Greek Science* (Harmondsworth: Penguin/Pelican, 1953) is clear, reasonably detailed, and informative. See also Collingwood, part I. The picture of ancient science receives an added dimension from the essays of some outstanding specialists collected in *Toward Modern Science* (ed. Robert M. Palter, New York, 1961), volume I.

A stimulating sketch of the Greek historical background is H. D. F. Kitto, *The Greeks* (Harmondsworth: Penguin, 1951–57).

Hellenistic science is the subject of a first classic in this entire pioneer field, George Sarton's *A History of Science: Hellenistic Science and Culture in the Last Three Centuries B.C.* (Cambridge: Harvard University Press, 1959). A brief clarification of the major Hellenistic concepts within a vividly re-created cultural setting is presented by Kenneth Heuer, *City of the Stargazers: The Rise and Fall of Ancient Alexandria* (New York, 1972).

On the rebirth of the concept of natural law in twelfth-century scientific thought, see the bibliographical note for Chapter Three. The idea of a mathematically structured universe seems to underlie rather clearly the physical speculations of mathematicians at Oxford's Merton College, e.g., William Hentisberus; see, e.g., Ernest A. Moody, "Laws of Motion in Medieval Physics," *Toward Modern Science*, volume I, pp. 226ff. For the same idea in the Renaissance, see Joan Gadol, *Leon Battista Alberti: Universal Man of the Early Renaissance* (Chicago: University of Chicago Press, 1969); also Ernst Cassirer, *The Individual and the Cosmos in Renaissance Philosophy* (trans. Mario Domandi, New York and Evanston: Harper Torchbooks, 1963). The world is likened to a tabernacle by Cosmas of Alexandria (sixth century) in his *Topographia Christiana*, as part of his argument against the ancient cosmographers; see Lloyd A. Brown, *Story of Maps* (Boston, 1950), pp. 89–90.

On the nature of early Christianity, in particular St. Augustine's philosophical contribution to the saving of Western civilization, the most perceptive discussion, to this writer's knowledge, is W. G. De Burgh, *The Legacy of the Ancient World* (Harmondsworth: Penguin/Pelican, 1955, originally published 1923), volume II, chapter 9, "Christianity" (pp. 374ff. on St. Augustine). Etienne Gilson's *The Spirit of Medieval Philosophy* (trans. A. H. C. Downes, New York, 1940) is a profoundly knowledgeable introduction to Medieval thought. In a much less exacting vein is Anne Fremantle's *The Age of Belief* (New York: Mentor, 1955), consisting of representative excerpts and insightful commentary. On the relationship between Medieval thought and nature, see especially Gilson, *Spirit*, chapter 18.

Books conveying a vigorous sense of the major historical forces at work in the Middle Ages are rare and seem to get rarer with the growth of specialized scholarship. Something of an exception is Christopher Dawson, *The Making of Europe* (Cleveland and New York: Meridian, 1956, originally published 1932). Excellent, especially for the later period, is Wallace K. Ferguson, *Europe in*

Transition: 1300–1520 (Boston, 1962), which projects a vivid picture of the social and economic scene during the High Middle Ages, including preceding developments. Substantial insights into the making of Medieval society from the viewpoint of technological progress are conveyed by Lynn White, Jr., *Medieval Technology and Social Change* (New York: Galaxy, 1966). Charles Homer Haskins' classic picture of the twelfth-century scene, *The Renaissance of the Twelfth Century* (New York: World–New American Library, 1957–1976, originally published 1927), has been updated in certain respects for more recent scholarship in *Twelfth Century Europe and the Foundations of Modern Society* (ed. Marshall Clagett, Gaines Post, and Robert Reynolds, Madison: University of Wisconsin Press, 1961).

St. Augustine's condemnation of the ancient sciences occurs in his *Enchiridion* (trans. J. F. Shaw, Edinburgh, 1872), chapter 9. Alfred North Whitehead's statement linking "faith in the possibility of science" with Medieval theology is part of a profoundly perceptive passage about modern science's essential Medieval heritage, in his *Science and the Modern World* (New York: Mentor, 1963, originally published 1925), p. 19. Roger Bacon's amazing prediction of modern technological inventions occurs in his *Epistola de secretis operibus*, trans. T. L. Davis as *Roger Bacon's Letter Concerning the Marvelous Power of Art and Nature and Concerning the Nullity of Magic* (Easton, Pennsylvania, 1923), pp. 26–27. Aquinas cautions against overestimating the powers of human reason in his *Summa contra gentiles* (trans. the Dominican Fathers, London, 1924), especially book I, chapter 8.

Chapter Three: Science and Faith at Chartres (pages 65–91)

On the cathedral of Chartres, its building history, and so on, see Emile Mâle, *Notre-Dame de Chartres* (Paris, 1948), especially chapter 2, "La cathédrale du XIIᵉ siècle"; also chapter 4, "Vue d'ensemble. La sculpture." For the School of Chartres, see G. Paré, A. Brunet, P. Tremblay, *La renaissance du XIIᵉ siècle: Les Ecoles et l'enseignement* (Paris–Ottawa: Publications de l'Institut d'Etudes Médiévales d'Ottawa, III, 1933, especially chapter 4, "La renaissance . . . Les scientifiques"). This is also useful for institutional detail, curricula, and such. See also A. Clerval, *Les écoles de Chartres au Moyen Age* (Paris, 1895).

For Chartres's humanist ideals and educational reforms, in particular under Thierry's chancellorship, see Raymond Klibansky's illuminating essay, "The School of Chartres," in *Twelfth Century Europe and the Foundations of Modern Society* (ed. Marshall Clagett, Gaines Post, and Robert Reynolds, Madison: University of Wisconsin Press, 1961). Also Tullio Gregory, *Anima Mundi: La filosofia di Guglielmo di Conches e la scuola di Chartres* (Florence, 1955), chapter 5, "Gli ideali culturali della Scuola di Chartres." M. D. Chenu's, O.P., "Nature and Man at the School of Chartres in the Twelfth Century," in *The Evolution of Science* (ed. Guy Métraux and François Crouzet, New York: Mentor, 1963), provides a brief, stimulating introduction to the School's general significance and thought. See also the articles "Chartres" in the *Enciclopedia Italiana* (1931–39) and "John of Salisbury" in the *Encyclopedia Britannica* (1957).

Chartres's natural philosophy within the historical context of Medieval thought is lucidly discussed by Etienne Gilson, *The Christian Philosophy in the Middle Ages* (New York, 1955), part IV, chapter 3, "Platonism in the 12th Century," and passim. Also his *Spirit of Medieval Philosophy* (trans. A. H. C. Downes, New York, 1940). Thumbnail summaries of the school's major philosophies, with careful listings of the pertinent monograph literature, are in Friedrich Ueberweg's *Grundriss der Geschichte der Philosophie* (11th edition, ed. Bernhard Geyer, Berlin, 1928), pp. 226ff. See also J. M. Parent, *La doctrine de la création dans l'Ecole de Chartres* (Paris 1938), with extensive excerpts from Thierry's and Conches's writings. Also Martin Grabmann, *Die Geschichte der Scholastischen Methode* (volume II: Die Scholastische Methode im 12. und beginnenden 13. Jahrhundert, Graz, 1957), part 2, chapter 6, "Die Schule von Chartres." For William of Conches, in addition to the aforementioned, I have mostly relied on Tullio Gregory's thorough study, including his copious excerpts.

John of Salisbury's praise for William of Conches occurs in the *Metalogicon* (trans., with comments, by Daniel D. McGarry, Berkeley: University of California Press, 1962), book I, chapter 5; and book II, chapter 10, pp. 21 and 97. Alexandre Koyré's otherwise highly stimulating *From the Closed World to the Infinite Universe* (New York: Harper Torchbooks, 1958) tends to perpetuate the misconception that Medieval science saw the universe as an essentially static and "closed cosmos." On Nicholas of Cusa's "dynamic" concept of the universe, see Koyré, pp. 8ff. For Giordano

Bruno's explicit assertion of an ever-expanding, infinite universe, see Ibid., pp. 43ff.

For St. Augustine's *"rationes seminales,"* see Gregory, *Anima Mundi,* pp. 179–180. The reference to the "discovery of nature" is in Chenu's essay, p. 223. Thierry explaining creation *"secundum physicam"* is in his *De sex dierum operibus,* p. 52; see Gregory, pp. 178, 182, n. 2. The version *"juxta physicas rationes tantum"* is quoted in Klibansky's essay, p. 8. See Gregory, pp. 177–178, 178, n. 2, 3, 181 on the concept of the beautification of the world (*exornatio mundi*). Gilbert's dismay over the "insanity of his time," as he calls it, and his feelings that a large segment of the students "would end up as bakers" is reported by John of Salisbury in the *Metalogicon,* book I, chapter 5, including the statement that Thierry was among those who tried to counteract this trend; p. 21.

William of Conches's "I take nothing away from God..." is here quoted from Chenu's essay, p. 227. See Gregory, p. 236, for the Latin (from Conches's *In Boetium,* ed. Parent, p. 126). For the scope of Conches's commentaries, including extant manuscript sources, see Klibansky's essay, pp. 10–11, in addition to Gregory. The problem of "second causes" (pervasive in Chartres's natural philosophers) is discussed by Gregory, pp. 175–176. For Conches's "like things ... by like things" (*Opus nature est quod similia nascantur ex similibus, ex semine vel ex germino, quia est natura vis rebus insita, similia de similibus operans*), see Gregory, pp. 178, n. 5, 179, n. 1. (The definition occurs in Conches's *Dragmaticon,* p. 31; and in his *In Timaeum,* ed. Parent, p. 147; and ed. Schmid, pp. 234–235.) Conches's "to seek the reason..." is quoted from Chenu, pp. 227–228. Adelard of Bath's paean to nature's rational beauty occurs in his *Quaestiones naturales* and is here quoted from a paraphrase by Chenu, p. 228. (See Gregory, p. 241, n. 1 for a lengthy passage from Adelard in praise of reason.)

John of Salisbury's frequent references to Ecclesiastes tend to multiply in the last part of the *Metalogicon,* increasingly tinging its mood (see pp. 258, 261, 269, 272).

Chapter Four: The Gift of Islam (pages 92–129)

The most informative work on Arab culture, including many detailed aspects of science, is Philip K. Hitti, *The History of the Arabs* (5th edition, New York, 1951, originally published 1937). A handy condensation by the same author is *The Arabs: A Short History* (Chicago: Gateway, 1943–49). H. A. R. Gibb, *Mohammedanism: An Historical Survey* (Oxford University Press, 1948–62), presents a similarly helpful introduction for the non-Muslim, with emphasis on the religious tradition. A stimulating historical survey is Bernard Lewis, *The Arabs in History* (New York: Harper Torchbooks, 1960, originally published 1950). Probably the best introduction to the impact of Islam on Medieval culture remains Charles Homer Haskins' classic *The Renaissance of the Twelfth Century* (New York: World–New American Library, 1957–1976). Virtually all the above shorter volumes serve as useful introductions to an alien civilization for the Western-educated. One might add the highly personal but captivating sketch by William Bolitho, "Mahomet," in his *Twelve Against the Gods* (New York: Modern Age Books, 1937, originally published 1929). But with the exception of Haskins and Hitti's *Short History*, all are notoriously thin on science, so that both for this and any other more detailed aspect, one has to refer continuously to the "big Hitti" or, specifically for science, to the more specialized histories of individual disciplines.

Medieval Europe's assimilation of Islamic science emerges from a combination of relatively general histories and more specialized studies, e.g.: A. Rupert Hall, *The Scientific Revolution, 1500–1800* (2nd edition, London, 1962), passim (see especially the index, "Islamic science"); A. C. Crombie, *Medieval and Early Modern Science* (New York: Doubleday Anchor, 1959), volume I, pp. 33ff.; and volume II, chapter 1, passim; Paul Oskar Kristeller, *The Classics and Renaissance Thought* (Cambridge: Harvard University Press, 1955, and New York: Harper Torchbooks, 1961), pp. 24ff., chapter "The Aristotelian Tradition," passim; Charles Singer, *A Short History of Science to the 19th Century* (Oxford, 1941–43), pp. 140ff.; I. Bernard Cohen, *The Birth of a New Physics* (New York: Anchor, 1960), especially chapters 2 and 3, concerning the assimilation of the Aristotelian and Ptolemaic systems; the articles "Arabic Achievement in Physics" by H. J. J. Winter, and "Medieval Medicine" by Henry E. Sigerist in *Toward Modern Science* (ed. Robert M. Palter, New York, 1961), volume I; and again the various histories of specialized disciplines (see below).

Islamic medicine and pharmacology are discussed in Haskins, pp. 322ff.; also pp. 52, 282–283, 290, 301; Hitti's *Short History*, pp. 141ff.; Charles Singer, *From Magic to Science* (New York: Dover, 1958, originally published 1928), pp. 188–189, 240ff.; Crombie, volume I, pp. 48ff., 223–224; Sigerist's "Medieval Medicine"; Thomas E. Keys, *The History of Surgical Anesthesia* (New York: Dover, 1963, originally published 1945), chapter "The Beginnings," passim; and the "big Hitti," pp. 363ff. See also A. M. Campbell, *Arabian Medicine and its Influence on the Middle Ages* (2 volumes, London, 1926). Islam's contribution to alchemy and chemistry is discussed by Haskins, pp. 319ff.; Hitti, *Short History*, pp. 147–148; David Pramer, "The Ancient Art of Alchemy," *Natural History*, August–September 1963; or, in its broader scientific context, by E. J. Holmyard, "Chemistry in Islam," in *Toward Modern Science*, volume I. On Arab astronomy (and astrology): Singer, *Short History*, pp. 151–152; Hitti, *Short History*, pp. 145ff.; W. W. Tarn, *Hellenistic Civilization* (Cleveland and New York: Meridian, revised edition, 1952), pp. 345ff., concerning ancient origins; "big Hitti," pp. 373ff. On al-Mamun's astronomers measuring the length of a degree: Hitti, *Short History*, p. 146; "big Hitti," pp. 568–569, 609–610. On Arab numerals, see below. On the manufacture of paper, "one of Islam's most beneficial contributions to Europe," and its introduction from Morocco into Spain (cf. the term "ream" from Arab *rizmah*, bundle): "big Hitti," pp. 564–565.

The "big Hitti," Ibid., conveys a lively picture of the role of the book in Islamic cultural life, as well as the fate of the Muslim libraries in Spanish hands. On the translations: Haskins, pp. 278ff.; "big Hitti," pp. 563ff; Kristeller, "The Aristotelian Tradition," pp. 30ff.; also his "Paduan Averroism and Alexandrism in the Light of Recent Studies," *Renaissance Thought II* (New York: Harper Torchbooks, 1965), for the larger intellectual context; Hitti, *Short History*, pp. 191–192; Holmyard, p. 168; Hall, p. 4; Crombie, volume I, pp. 33ff., including helpful chronological tables. On the translations by Renaissance humanists: Crombie, volume II, pp. 104ff. For the stir created by Aristotle's scientific writings, see Helene Wieruczowski, *The Medieval University* (Princeton: Anvil, 1966), pp. 41–42. Kristeller, "Aristotelian Tradition," pp. 30ff., describes the reception of Aristotle's philosophy in the Medieval West.

For al-Edrisi's background and work, see "big Hitti," pp. 609–610. On Frederick II's youth, Arab influences, scientific interests, etc., see Ernst Kantorowicz, *Frederick the Second* (trans. E. O. Lorimer, New York, 1957, German original, 1931), passim. For his relationship with Michael Scot: Ibid., pp. 339–340, and passim. For his falcon book (*De arte venandi cum avibus*): Ibid., pp. 359ff. On Michael Scot see also Lynn Thorndike, *Michael Scot* (Lon-

don, 1965), especially chapter 3, "Michael Scot as a Translator."
On the historic role of the Nestorians at Gundeshapur (Jundi-
shapur) and other Persian centers of learning: William Cecil Dampier, *A Shorter History of Science* (New York: Meridian, 1957), pp.
37–38; Hitti, *Short History*, p. 118; Haskins, p. 281; Crombie, volume I, p. 33. Islamic observatories (including the role of Omar-
Khayyám): "big Hitti," pp. 377–378. On Islamic physics, see Winter's "Arabic Achievement in Physics," above. See Ibid., pp. 172ff.
on Alhazen; also Crombie, volume I, pp. 49, 101–102, and Hitti,
Short History, pp. 145–146, on al-Kindi.

Islam's contribution to mathematics is lucidly discussed by
W. W. Rouse Ball, *A Short Account of the History of Mathematics*
(New York: Dover, 1960, originally published 1908), pp. 144ff.,
chapter 9, "The Mathematics of the Arabs." Ibid., pp. 121ff., chapter 7, "Systems of Numeration and Primitive Arithmetic," on the
general evolution of numerals. For the Hindu contribution: the
chapter "The Hindus" from Florian Cajori's *A History of Mathematics* (2nd edition, New York, 1919, reprinted in *The Growth of
Mathematics*, ed. Robert W. Marks, New York: Bantam, 1964), pp.
91ff. Some additional aspects of Islamic mathematical developments are given by D. E. Smith, *History of Mathematics* (New
York: Dover, 1958, originally published 1923), volume I, pp. 164ff.
Some revealing insights, especially about al-Khwarizmi's achievement and its impact on the Medieval West, are offered by Egmont
Colerus, *Von Pythagoras bis Hilbert* (Berlin–Vienna, 1937), chap-
ters 6, "Alchwarizmi," and 7, "Leonardo von Pisa." Also Hitti,
Short History, pp. 183–184.

Some important qualifications concerning the role of Averroism
in the light of recent studies are noted in P. O. Kristeller's essay
(see above). For a brief summary of Averroës' thought, see, e.g.,
Frederick B. Artz, *The Mind of the Middle Ages* (3rd edition, New
York, 1959), pp. 160–161. For Aquinas' tangling with the "Latin
Averroists" in Paris, his involvement in the subsequent backlash,
and Albert's posthumous rescue mission, see Anne Fremantle, *Age
of Belief* (New York: Mentor, 1955), pp. 144–145, 149; Wieruczowski, *Medieval University*, pp. 45ff., especially p. 47. Condemnation
of Siger's and related "errors": Ibid., p. 50. For Averroism in later
Medieval and Renaissance science, see John Herman Randall, Jr.,
The School of Padua and the Emergence of Modern Science (Padua, 1961). On the banning of Aristotle's scientific writings in
Paris and their gradual emergence in the classrooms: Wieruczowski, pp. 41–42; on their teaching at Toulouse: Ibid., pp. 41, 84–85,

179. The slow adoption of Arab numerals is described by Crombie, volume I, pp. 51–52. Leonardo Fibonacci (Leonardo of Pisa), his *Liber abaci,* and the gradual penetration of Arab arithmetic into European science and trade are discussed by Rouse Ball, pp. 167ff.; Colerus, pp. 138ff.

About Renaissance humanists dabbling in science, see, e.g., P. O. Kristeller, "The Platonic Academy of Florence," *Renaissance Thought II,* p. 91, concerning Ficino. (Actually, Ficino's writings abound in scientific references, usually in the guise of Aristotelian concepts.) Islamic influences on Medieval chivalry, poetry, and art are discussed, e.g., in Denis de Rougemont, *Love in the Western World* (New York: Doubleday Anchor, 1957, originally published 1940), passim; Hitti, *Short History,* pp. 174ff. (See Ibid., pp. 182–183, 186 on European terms taken over from Arabic.) Kantorowicz, *Frederick the Second,* pp. 283–284, on the introduction of the Arab customs system.

Roger Bacon's sneer at the translators occurs in his *Opus tertium,* chapter 25 (*Opera inedita,* ed. J. S. Brewer, London, 1859, p. 91). The passage — scathing in its entirety — was brought to my attention by my late, revered friend Helene Wieruczowski; it is now included in her *Medieval University,* p. 161. Sarton compares Euclid's *Elements* with the Parthenon in his *History of Science: Hellenistic Science and Culture in the Last Three Centuries B.C.* (Cambridge: Harvard University Press, 1959), p. 39.

Dante places Averroës in the Limbo of the unbaptized and the virtuous heathen in the Inferno, Canto IV, l. 144. The masters of Toulouse invited students to attend the university in a circular letter in 1229, including the sentence: "Whoever wants to scrutinize the bosom of nature can hear lectures on the *libri naturales,* the books that have been forbidden at Paris..." (Wieruczowski, pp. 178–179). The quotation at the head of the chapter is taken from *The Meaning of the Glorious Koran,* translated, with comments, by Mohammed Marmaduke Pickthall (New York: Mentor, 1953–54).

Chapter Five: Scholastics, Mystics, Alchemists (pages 130–187)

Technological developments, particularly during the later Middle Ages, are masterfully presented in Lynn White, Jr.'s, *Medieval Technology and Social Change* (New York: Galaxy, 1966). R. G.

Collingwood, in the introduction to *The Idea of Nature* (New York: Galaxy, 1960), suggests something of the interdependence between science and the historical cultures that have furthered its growth. The contribution of Medieval mysticism to science is most fully treated in Lynn Thorndike, *A History of Magic and Experimental Science During the First Thirteen Centuries of Our Era*, volumes I and II (New York, 1923). See also Charles Singer's *From Magic to Science* (New York: Dover, 1958), passim; E. A. Burtt, *The Metaphysical Foundations of Modern Science* (New York: Doubleday Anchor, n.d., originally published 1924), passim; and Alfred North Whitehead, *Science and the Modern World* (New York: Mentor, 1963), passim. However, the substantial positive contribution of Medieval mysticism to science seems nowhere adequately acknowledged in the historical literature except for some specialized histories of the individual "magic" sciences (see below). (Thorndike's formidable scholarship, manifest throughout his study, seems in this respect somewhat distorted by his rationalist bias, so that the mystic influences behind Medieval science by and large come through as little more than a cultural curio shop.)

Medieval rationalism is discussed, besides the studies by Gilson, Ueberweg, and Grabmann mentioned in the bibliographic note for Chapter Three, e.g., in Etienne Gilson, *Reason and Revelation in the Middle Ages* (New York: Scribner Library, n.d., originally published 1938); also Anne Fremantle's *Age of Belief* (New York: Mentor, 1955), and F. C. Copleston's handy — if not very stimulating — survey, *Medieval Philosophy* (New York, 1961, originally published 1952). The early theological disputes are treated in a meaningful historical context in W. G. De Burgh's *Legacy of the Ancient World* (Harmondsworth: Penguin/Pelican, 1955), volume II, chapter 9, "Christianity." On Roger Bacon's quintessential mysticism: Steward C. Easton, *Roger Bacon and His Search for a Universal Science* (New York: Columbia University Press, 1952), pp. 73ff.

The magic inconsistencies of Medieval scientists are richly documented, e.g., for Albertus Magnus, by Thorndike, *Magic*, volume II, pp. 548ff. For astrological beliefs among humanists, see e.g., P. O. Kristeller, "The Moral Thought of Renaissance Humanism," *Renaissance Thought II* (New York: Harper Torchbooks, 1965), p. 58. Petrarch ridicules the academic snobbishness of the Scholastic tradition (including some of its more abstruse scientific ideas) in, e.g., his treatise "On His Own Ignorance and That of Others," *The Renaissance Philosophy of Man*, ed. Ernst Cassirer, P. O.

Kristeller, John Herman Randall, Jr. (Chicago: University of Chicago Press, Phoenix Books, 1948–58), pp. 47ff. See also the excellent introduction and comments by Hans Nachod.

For the major dynamics of Medieval civilization, see the bibliographic note for Chapter Two. A sense of the fundamental tensions underlying high Medieval culture is reflected in Richard W. Southern, *The Making of the Middle Ages* (London, 1953), passim; and *Medieval Humanism and Other Studies* (Oxford, 1970), especially the title essay. A profound understanding for the meaning of the mystic vision is conveyed by Evelyn Underhill, *Mysticism: A Study in the Nature and Development of Mysticism's Spiritual Consciousness* (10th edition, London, 1923), especially her sampling of representative quotations, pp. 231–232. On the creative element in science, see, e.g., Anne Roe, "The Psychology of the Scientist," in *The New Scientist: Essays on the Methods and Values of Modern Science,* ed. Paul C. Obler and Herman A. Estrin (New York: Doubleday Anchor, 1962).

An open-minded attitude toward nonrationalist approaches to science pervades Alfred North Whitehead's *Science and the Modern World;* also E. A. Burtt's *Metaphysical Foundations of Modern Science* (with both studies predicated on a critique of the philosophical premises of modern science). See also the balanced discussion by L. Arnaud Reid, "Religion, Science and Other Modes of Knowledge" in *The New Scientist.* An idea of the inadequacy of accepted scientific methods before the problems facing contemporary science emerges from Erwin Schrödinger's famous essay, "What is Life?" in his *What is Life? and Other Scientific Essays* (New York: Doubleday Anchor, 1956).

For Albertus Magnus' observations of whaling, fishing, etc., as Dominican provincial for Germany, see A. C. Crombie, *Medieval and Early Modern Science* (New York: Doubleday Anchor, 1959), volume I, p. 142. For more extensive comments on his zoological and botanical observations: Ibid., pp. 147ff. For Albertus on therapeutic properties of various lion's parts, see Thorndike, volume II, pp. 660–661. For Newton's alchemical interests, see E. N. da C. Andrade, *Sir Isaac Newton: His Life and Work* (New York: Doubleday Anchor, n.d., originally published 1954), pp. 129ff. See Ibid., pp. 130, 132, on his alchemical exchanges with Robert Boyle. On the historic continuity from alchemy to chemistry, see J. R. Partington, *A Short History of Chemistry* (London, 1937). See also E. J. Holmyard, *Makers of Chemistry* (Oxford, 1931), and *Alchemy* (London: Pelican, 1957).

For Copernicus' frequent references to the Alfonsine Tables, see *Three Copernican Treatises*, ed. Edward Rosen (New York: Dover, 1959, originally published 1939), index, "Alfonsine Tables." In publishing a first account of Copernicus' theory, the *Narratio prima*, Rheticus emphasized that Copernicus had devised his theory of a moving earth in order to account for the observational evidence "of all ages" — i.e., that had been amassed from antiquity until his own time: Ibid., pp. 110, 136–137. There is a considerable historical literature on the so-called problem of saving the appearances or, in other words, the pressures of the expanding observational evidence on the evolution of astronomical theory. For a concise summary, see John Herman Randall, Jr., *The Making of the Modern Mind* (Boston: Houghton Mifflin Co., revised edition, 1940), pp. 227ff. For a fuller discussion of the conceptual context, see Herbert Butterfield, *The Origins of Modern Science* (New York: Collier, 1962, originally published 1957), chapter 2, "The Conservatism of Copernicus"; also T. Goldstein, "The Renaissance Concept of the Earth in its Influence upon Copernicus," *Terrae Incognitae*, volume IV, 1972.

Albertus Magnus' attitude toward astrology seems to have been ambivalent throughout: at one point in his *Summa theologiae* he mustered the standard arguments against astrology from the Church Fathers to the contemporary rational objections. (See Thorndike, *Magic*, volume II, pp. 591–592, who however professes to doubt the authenticity of the passage.) He also denied that the stars affect the human free will or soul (Ibid., p. 584), condemned the "sophistry" of divinations (Ibid., volume I, p. 359), and denounced the "delight[ing] in deception" of the great majority of practitioners of astrology and related arts (Ibid., volume II, p. 585). But, as Thorndike emphasizes, he generally accepted the principles of astrology and on occasion defended them against rationalist attack (Ibid., volume II, pp. 577ff.).

On Medieval medicine, see Henry E. Sigerist, "Medieval Medicine," *Toward Modern Science* (ed. Robert M. Palter, New York, 1961), volume I; Crombie, volume I, pp. 222ff.; also pp. 168ff.; Singer, *From Magic to Science*, chapters "Early English Magic and Medicine," "Early Herbals," and "The School of Salerno and Its Legends"; Thorndike, *Magic*, volume I, chapter 31, "Anglo-Saxon, Salernitan, and other Latin Medicine in Manuscripts from the Ninth to the Twelfth Century," and Ibid., volume II, passim. See also Madeleine Pelner Cosman, "Machot's Medical-Musical

World," in *Machot's World: Science and Art in the Fourteenth Century* (ed. M. P. Cosman, New York: New York Academy of Science, 1978). The origin of the Medici coat of arms is actually disputed, but for the likely roots in the apothecary guild see, e.g., Hugh Ross Williamson, *Lorenzo the Magnificent* (New York, 1974), pp. 29–30. Raymond deRoover, *The Rise and Decline of the Medici Bank, 1397–1494* (New York: Norton Library, 1966, originally published 1963), p. 15, categorically denies the connection between the pawnbrokers' symbol and the Medici coat of arms (see also his "The Three Golden Balls of the Pawnbrokers," *Bulletin of the Business Historical Society*, XX, 1946).

For Salerno's medical school and the *Regimen sanitatis Salernitanum*, see Wieruczowski, chapter 6, "Salerno and Montpellier: The Study of Medicine," including Salerno's fame as "the city of Hippocrates" (Ibid., p. 77), with samples from the *Regimen* (Ibid., pp. 174–175). Also Crombie, volume I, pp. 168–169, 180–181, 223–224. For Constantine the African, see Wieruczowski, pp. 75–76; Crombie, volume I, pp. 34, 223–224; more detail in Thorndike, *Magic*, volume I, chapter 32. (See also Singer, *From Magic to Science*, pp. 244–245, who seeks to dispute Constantine's merits.) On herbals, see the excellent chapter "Early Herbals" in Singer, *From Magic to Science*; Crombie, volume I, pp. 145ff.; also Thorndike, *Magic*, volumes I and II, passim (index, "herb," "herbs," and "herbals").

For the *Macer floridus*, and *Circa instans*, see Crombie, volume I, p. 146; Singer, *From Magic to Science*, pp. 74, 188–189. On Albertus Magnus' *De vegetabilibus et plantis*, see Crombie, volume I, pp. 147ff. For Albertus on the virtues of plants, see Thorndike, *Magic*, volume II, pp. 564ff. For Albertus deferring to the magi for a more detailed knowledge of the subject, Ibid., pp. 555–556. For Albertus on astrological influences on plants, Ibid., pp. 564ff. For Albertus referred to as "*Magnus in magia*," see Ibid., p. 554. On Albertus' personality and defense of Aquinas, see the beautiful chapter "Albertus Magnus" in Henry Osborn Taylor's *The Medieval Mind* (Cambridge: Harvard University Press, 1959, originally published 1911), volume II.

For use of anesthetic sponges, see Thomas E. Keys, *History of Surgical Anaesthesia* (New York: Dover, 1963), pp. 7ff. For Michael Scot's prescription, see Crombie, volume I, p. 225. (A more explicit twelfth-century recipe, from the *Antidotarium* of Nicolas of Salerno, is cited in Keys, p. 7.) For earliest botanical gardens, see

Crombie, volume I, p. 158. For evolution of botanical illustrations into art, see Singer, *Magic*, pp. 190ff.

A fine feeling for the interplay of engineering and esthetic effect in Gothic architecture is conveyed by E. H. Gombrich, *The Story of Art* (New York, 1955), pp. 131ff. For a more technical explanation of the problem of the pointed arch, see Crombie, volume I, pp. 203ff. Gothic technology is discussed with much attractive detail by Jean Gimpel, *The Cathedral Builders* (trans. Carl F. Barnes, Jr., New York: Evergreen, 1961). For heights of individual cathedrals, see Ibid., p. 6. On the "Gothic crusade": Although the wave of collective enthusiasm and active popular participation accompanying the rise of the early cathedrals is impressively attested to by contemporary sources (e.g., the prior of Mont Saint-Michel reporting on the building of Chartres cathedral in 1145, quoted in *French Cathedrals*, introduction by François Fosca, trans. Dorothy Plummer, Munich, 1959, pp. 6–7), some art historians emphatically reject this notion: e.g., Emile Mâle, *Religious Art From the 12th to the 18th Century* (New York: Noonday, 1958–59, originally published 1949), p. 29.

On upward motion in Medieval mysticism, see e.g., Anne Fremantle's *Age of Belief*, chapter on St. Bonaventura, including excerpts from his *Journey of the Mind to God*. The adaptation of the anagogical approach to architectural principles is discussed by Erwin Panofsky, "Abbot Suger of St. Denis," reprinted in his *Meaning in the Visual Arts* (New York: Doubleday Anchor, 1955), pp. 128ff. Concerning the role of light in the mystic tradition and the Gothic style, see Ibid., pp. 127ff. On Suger's passion for sumptuous decoration, see Gimpel, pp. 23ff. For Isidore of Seville and Honorius of Autun as sources for Gothic monsters, see Emile Mâle, *The Gothic Image* (trans. Dora Nussey, New York: Harper Torchbooks, 1958, originally published 1913), pp. 39ff., concerning Honorius' *Speculum Ecclesiae*. See Mâle, *Religious Art*, pp. 44ff., on both Isidore and Honorius.

For Gothic décor as a replica of the surrounding flora, see the delightful passage in Mâle, *Gothic Image*, pp. 52ff. The perception that this floral décor might mirror the advance of the seasons (Ibid., p. 52) is based on Viollet-le-Duc's article "Flore" in his *Dictionnaire raisonné de l'architecture française du XIᵉ au XVIᵉ siècle*. For Masons' and position marks, see Gimpel, pp. 85ff. On the masons and their lodges, see Ibid., pp. 91ff. For multiple meanings in Medieval art, Emile Mâle's *Gothic Image* presents telling (and now famous) examples; see especially his introduction. For the

layout and building history of Siena, see e.g., the little volume *Siena* by Paolo Cesarini (trans. Googie Maraventano, Siena, n.d.).
Medieval alchemy is attractively treated in M. Caron and S. Hutin, *The Alchemists* (trans. Helen R. Lane, New York: Evergreen, 1961). The little volume includes most of the relevant aspects. See also Crombie, volume I, pp. 129ff.; E. J. Holmyard, "Chemistry is Islam," *Toward Modern Science*, volume I; Easton, *Roger Bacon*; also Thorndike, *Magic*, volume II, chapters 45 and 65 (and passim); and the bibliography for Chapter Four, above. For Nicholas Flamel's story, see Caron and Hutin, pp. 6ff. For equipment of an alchemist's shop, see Ibid., pp. 62ff., chapter "The Laboratory." For the role of gold, see Ibid., pp. 75ff., chapter "The Gold Mirage: The 'Puffers,' " and pp. 70ff., chapter "The Philosoper's Stone," and the above-named literature. Roger Bacon's prophetic statement about the scientific method occurs in chapter 11 of his *Opus tertium*, and is reprinted (in English) in Helene Wieruczowski, *The Medieval University* (Princeton: Anvil, 1966), p. 162.

Newton's passage about the Hermetic writers is from one of his letters and refers to the possibility of transmuting metals to gold. ("[It] has been thought fit," the complete passages reads, "to be concealed by others that have known it, and therefore may possibly be an inlet to something more noble, not be be communicated without immense danger to the world . . ."; quoted in E. N. da C. Andrade, *Sir Isaac Newton, His Life and Work*, pp. 132–133.) Albertus Magnus' encomium for the manifold virtues of the nasturtium is taken from his *De vegetabilibus et plantis*, VI, ii, 13, quoted by Thorndike, *Magic*, volume II, p. 565. His attribution of healing powers of herbs to their closeness to the ground is from *De vegetabilibus*, VI, ii, 1, quoted by Thorndike, p. 564. Abbot Suger's inscription at the choir of St. Denis is here quoted from Gimpel, *Cathedral Builders*, pp. 26–27. The Latin original (with a somewhat different translation) is in Panofsky, "Abbot Suger," *Meaning*, pp. 129–130. Suger's invocation of the Holy Martyrs as sponsors of his sumptuous altar is from his memorial, written after 1127 (see Panofsky, pp. 122ff.) and quoted here from Gimpel, p. 25. The distinction between "false alchemists" and "true philosophers," quoted by Gimpel, p. 79, is from the *Physica subterranea* of Johannes Joachim Becher, a seventeenth-century author (Leipzig, 1738), p. 25.

Chapter Six: Art and Science in the Renaissance* (pages 188–241)

On city planning and urban expansion in Florence since the late thirteenth century, see Gene A. Brucker, *Renaissance Florence* (New York: 1969), pp. 25ff. (27ff. on actual city planning); the Medici family's role in this field, see pp. 34ff.; also Ferdinand Schevill, *The Medici* (New York: Harper Torchbooks), pp. 91ff. and passim. (Brucker's discussion is generally based on the best available scholarship, but Schevill's, though providing a useful overall picture, is in many ways outdated.) For the citizens' participation in planning for the cupola of the *duomo*, see Brucker, pp. 32–33, 217. Significant corrections and qualifications concerning the Medici role have been presented by E. H. Gombrich, "The Early Medici as Patrons of Art," in his *Norm and Form: Studies in the Art of the Renaissance* (London–New York, 1972, originally 1966.) For sixteenth-century additions to the layout of Florence, see Eric Cochrane, *Florence in the Forgotten Centuries* (Chicago: University of Chicago Press, 1973), pp. 90–91 and passim. The problems of the relationship between the Middle Ages and Renaissance — a subject of intense debate in recent decades — is handily summarized in *The Renaissance: Medieval or Modern?* (ed. Karl H. Dannenfeldt, Boston: D. C. Heath, 1959, part of the College series "Problems in European Civilization").

On the continuity of Medieval and early modern science, see Crombie, volume II, pp. 103ff., "The Continuity of Medieval and 17th Century Science." For Botticelli's and Leonardo's flower studies, see Charles Singer, *From Magic to Science* (New York: Dover, 1958), pp. 191–192. See also Kenneth Clark's perceptive discussion, *Leonardo da Vinci: An Account of His Development as an Artist* (Harmondsworth: Penguin, 1959, revised edition), p. 114.

For the gradual focusing of art upon nature, see, e.g., Kenneth Clark, *Landscape Into Art* (Harmondsworth: Penguin/Pelican, 1949–56). The problems resulting from the new program (including the training of young artists) are given special attention in E. H. Gombrich's *Story of Art* (New York, 1955), pp. 177ff. (chapter "Tradition and Innovation"), p. 179 (artist's training), 179ff., 190ff. (specific problems resulting from the new program). Renaissance theories about the rendering of human proportions are discussed with scholarly thoroughness (and a remarkable feeling for the relations between art and science) in Erwin Panofsky's "The His-

---

* Since Chapter Six is rather closely related to my current work, of which it represents some generalized conclusions, I have frequently set forth these conclusions as my own, without the usual supporting references.

tory of the Theory of Human Proportions as a Reflection of the History of Styles," *Meaning in the Visual Arts* (New York: Mentor, 1955), pp. 88ff.; see especially Ibid., pp. 89–90, n. 63, on the influence of Alhazen's *Optics* on Ghiberti's esthetic theories.

Gombrich, *Story of Art,* pp. 144ff. (Giotto) and p. 165 (Masaccio's *Trinity*) conveys a keen sense for the spatial innovations of the "new art." Of the extensive specialized literature, see, e.g., John White, *The Birth and Rebirth of Pictorial Space* (2nd edition, Boston, 1967), including the bibliography. Some thoughts about the Renaissance versus the Medieval concept of physical space are suggested in my "Role of the Italian Merchant Class in Renaissance and Discoveries," *Terrae Incognitae,* volume VIII, 1976, especially pp. 20–21. Specifically regarding the concept in science, see my article "The Renaissance Concept of the Earth in its Influence upon Copernicus," Ibid., volume IV, 1972, passim, especially n.18. (See Ibid. for the projection of terrestrial physics into universal cosmology, under the influence of the geographic discoveries.) For the dating of Giotto's Padua frescoes, see *Giotto: The Arena Chapel Frescoes* (ed. James Stubblebine, New York, 1969), p. 74. For the beginning of mathematical criticism of Aristotelian physics at Merton College (by Thomas Bradwardine, in his *Tractatus de proportionibus velocitatum in motibus,* 1328), see Ernest A. Moody, "Laws of Motion in Medieval Physics," *Toward Modern Science* (ed. Robert M. Palter, New York, 1961), volume I, pp. 224ff. Masaccio's St. Peter is walking through a Florentine street in his *St. Peter Healing the Sick,* circa 1426–27, Brancacci Chapel (Santa Maria del Carmine), Florence. (Masaccio frequently used Florentine backdrops in his frescoes.)

The relationship of motion in Renaissance art and the new geographic consciousness is suggested in my "Role of the Italian Merchant Class in Renaissance and Discoveries," p. 25. On the role of Renaissance art in scientific illustrations, see Crombie, volume II, pp. 264–265, 269–270, 272, 277; see also below concerning Vesalius. For Leonardo's anatomical studies I have mostly relied on the impressive volume, *Leonardo da Vinci On the Human Body* (ed. Charles D. O'Malley and J. B. de C. M. Saunders, New York, 1952), and its outstanding illustrations. On Vesalius, see Crombie, volume II, pp. 273ff. The affinities between the Ptolemaic map projection and the Renaissance theory of perspective are searchingly set forth by Joan Gadol, *Leon Battista Alberti: Universal Man of the Early Renaissance* (Chicago: University of Chicago Press, 1969), pp. 70ff. and passim. The inner connections between Leonardo's landscapes and maps have been noted by Kenneth Clark, *Leonardo da Vinci,* pp. 113–114. For the Mercator projection, see Lloyd A. Brown, *The Story of Maps* (Boston, 1950), pp. 135–136.

Leonardo's "Windsor" maps are discussed by Frederick Hartt, *Italian Renaissance Art* (Englewood Cliffs, New Jersey, 1969?), p. 389. Botticelli did an *Adoration of the Magi* in the early 1470s and another after 1482. Leonardo began his *Adoration* in 1481. The intrusion of exotic landscapes on the European imagination, including maps, is perceptively treated, with reference to the West Indies, by Hugh Honour, *The New Golden Land: European Images of America from the Discoveries to the Present Time* (New York: Pantheon, 1975), chapter 1, "First Impressions." The rivalry between Brunelleschi and Ghiberti is reported by Vasari (*Lives of the Artists,* abridged and edited by Betty Burroughs, New York, 1946), chapter "Filippo Brunelleschi," pp. 77ff. If the story is, in fact, at least partially apocryphal (as are, presumably, a good many of Vasari's stories), it seems noteworthy that Brunelleschi's successful plan for the construction of the cupola is related, in considerable detail, in the same context (Ibid., pp. 73ff.; Brucker, *Renaissance Florence,* p. 33).

Michelangelo's architectural works referred to here include his fountain in the cloister of San Marco, his staircase in the Laurenziana Library (as well as the library itself), and a number of façades or other structures for which Michelangelo either made the original design (sometimes modified later, like the Palace of the Senators on Capitol Hill in Rome), or appears to have added some characteristic touches. See, e.g., the chapter "Michelangelo" in Peter Murray, *The Architecture of the Italian Renaissance* (New York: Schocken, 1965–70). Among his architectural works one should also list the Medici Chapel (New Sacristy) in San Lorenzo, Florence, and the Porta Pia in Rome. For Vasari's story of the *David's* being hauled from the workshop to the Piazza della Signoria, see Vasari's *Lives of the Artists,* p. 262. For Michelangelo's using a new method for erecting a scaffold for the ceiling of the Sistine Chapel, see Ibid., p. 267. For his insistence on being a sculptor, not a painter, see Gombrich, *Story of Art,* p. 223.

See Joan Gadol's *Leon Battista Alberti* for Alberti's universality. For his contribution to cryptography, see Ibid., pp. 207ff. For his Pythagorean-Platonic theory of art as an embodiment of a natural order, see Ibid., pp. 104ff.; also pp. 150ff. (though there is some confusion on the cosmological context); and passim. For Alberti's contribution to perspective, see Ibid., chapter 1 (with extensive bibliography). On engraving (and its initial connection with printing), see Gombrich, Story of Art, pp. 204ff. Concerning Toscanelli acting as an informal consultant to the artists, see, e.g., Vasari's chapter on Brunelleschi, p. 71. (Toscanelli specifically acquainted the Florentines with a late fourteenth-century theory on perspective developed at the University of Padua, i.e., Biagio Pelacani's Quaestiones perspectivae; see Gadol, p. 27.) Alberti describes his "veil" (velo) — for which he on occasion used an actual reticulated cloth — in his Della pittura (ed. Luigi Mallè, Florence, 1950), pp. 83–84 (quoted in Gadol, p. 38). See Gadol, p. 39, n. 23, for the concept of the windowpane or "glass." Theories of perspective before Alberti (including Witelo's, Peckham's, and Roger Bacon's) are discussed by Gadol, pp. 22ff. Taddeo Gaddi's Presentation of the Virgin (Santa Croce, Florence), with its plethora of focal points, is reproduced in Gadol, fig. 5. On the perspective in Masaccio's Trinity and its effect on the public, see Gombrich, Story of Art, p. 165; Gadol, p. 23. For Brunelleschi's "centric point" scheme (or construction"), see Gadol, pp. 31–32, 35. On the "bifocal rule of construction," see Ibid., pp. 34–35. The precise nature of Alberti's innovation is lucidly discussed by Gadol, pp. 37ff. Alberti's specific mathematical contribution, based on his recognition of the basic affinity between the problems of perspective representation and of surveying (see also the section on mapping and perspective, above), is most clearly summarized in Gadol, pp. 40–41.

A comparatively short list of books on Leonardo (out of the nearly limitless bibliography) is in Kenneth Clark's Leonardo da Vinci, pp. 167ff. I would add to that the brilliantly thorough study by V. P. Zubov, Leonardo da Vinci (Moscow–Leningrad, 1962, and trans. David H. Kraus, Cambridge: Harvard University Press, 1968). A brief but vivid sketch of Leonardo's personality is in Pamela Taylor's introduction to her selection from The Notebooks of Leonardo da Vinci (New York: Mentor, 1960), pp. ix–x. For Leonardo's influence on Vesalius, see Crombie, volume II, pp. 274–275. (Crombie tends to credit Titian with the crucial influence on the illustrations of the De Fabrica.) Leonardo's plans for an anatomy are discussed in Leonardo da Vinci on the Human Body (ed. Charles D. O'Malley and J. B. de C. M. Saunders, New York, 1952), introduction, p. 31. For his more comprehensive project of a Trattato della pittura, see Clark, Leonardo, pp. 72ff. See also the probing chapter "The Notebooks," Ibid., pp. 60ff. For Leonardo's personal appearance, see Pamela Taylor, p. xiv. On the accusation of homosexuality (made twice in 1476), see Ibid., p. xiii. For a fuller, sensitive assessment see Clark, *Leonardo,* pp. 58–59. Leonardo's *Notebooks* are frequently interspersed with reminders to himself to pick up or borrow this or that book relating to a subject that currently fascinated him; see Zubov, passim. See also Clark, *Leonardo,* pp. 62–63, for an evaluation of his self-education. On Leonardo's refusal of other commissions during his sessions for the *Mona Lisa,* see Ibid., p. 112. The painting's "dream landscape" background is sensitively analyzed by Gombrich, *Story of Art,* pp. 219–220.

Leonardo's topical questions occur throughout his "Notebooks." See *The Notebooks of Leonardo da Vinci* (ed. Jean Paul Richter, New York: Dover, 1970, originally published 1883), 2 volumes; or the handy little volume edited by Pamela Taylor. (Man getting up from the ground, see Richter, volume I, no. 370; origin of rivers, see Taylor, p. 136; observing rock structure from inside a cave, see Ibid., p. 122; fish propelling itself by the tail, see Ibid., p. 122.) Leonardo's questions about light and shade are part of what has been judged a "complete and finished treatise on Light and Shade" (see Richter, p. 67; arranged in 6 books, Ibid., pp. 69ff). In fact, Richter's entire first volume deals with subjects of importance to the painter.

Aquinas' argument about "motion" is part of his five ways of demonstrating the existence of God, *Summa Theologica,* Q.2, third article, reprinted in Fremantle, *Age of Belief* (New York: Mentor, 1955), pp. 152–153. However, "motion" for Aquinas has a broader meaning, signifying at the same time something like evolution. A handsome analysis of Leonardo's first Arno Valley landscape is in A. Richard Turner, *The Vision of Landscape in the Renaissance* (Princeton: Princeton University Press, 1966), pp. 16ff. Turner, p. 16, also refers to the "master's graphic shorthand." Leonardo's *Notebooks* contain such startling cosmological statements as: "The earth is not the center of the Sun's orbit nor at the center of the universe . . ." and "the earth is a star . . ." See Richter, volume II, pp. 137, 139, notes 858, 865.

For Leonardo's reference to the eye as "King [or Lord] of the senses," see the quotation from the *Notebooks* in *The Age of Adventure: The Renaissance Philosophers* (ed. Giorgio de Santillana, New York: Mentor, 1956), p. 67. Boccaccio's praise of Giotto is reprinted in *Giotto in Perspective* (ed. Laurie Schneider, Englewood Cliffs, New Jersey: Spectrum, 1974), p. 28. (The quotation is from the *Decameron,* sixth day, fifth story.) Cennini's similar tribute is quoted from Schneider, p. 36 (from *The Art of the Old Masters as Told by Cennino Cennini*, trans. Christiana J. Herringham, London, 1899, p. 5). Leonardo speaks of the "fear of living . . . in the company of those corpses" in the introduction to his anatomical books (Taylor, p. 110). The sentence "I wish to work miracles" opens the section (Ibid.). The sentence describing the sequence of anatomical studies, in effect, in terms of a slow-motion movie occurs in the same context (Ibid.). His "The motive power is the cause of all life" is quoted from Taylor, p. 196; the sentence "Nature is full of infinite causes" and so on, from Ibid., p. 197.

Index

Index

Gnomon, 1n
Gold Makers' Lane (Slatá Ulička, (Prague), 178n, 179 (illus.)
Gothic cathedrals, 159 (illus.), 166 (illus.), 168 (illus.); phenomenon of, 156–62, 164–69; stained-glass windows, 162–64, 163 (illus.)
Gozzoli, Benozzo, The Visit of the Magi, 31, 32 (illus.)
Greece: intellectual freedom, 46; beginning of science, 46–47, 48–51; cultural development, 47–48, 50; attitude toward illness and medicine, 102–3; mathematical thoughts in, 117–18, 124
Grosseteste, Robert, 171n, 191, 192n

Hammurabi's Code, 118
Harmony, as ultimate aim of the soul, 53–54
Heliocentric system (sun-centered cosmos), 2, 131, 142, 199
Hemorrhoids, extirpation in Greek and Islamic medical practice, 101 (illus.)
Henry, Prince, the Navigator, 23, 40, 219
Herbals, use in Medieval medicine, 148–50, 156
Hermes Trismegistus, 179n, 181
Hermetic arts, 178–79
Hippocrates, 102, 145, 148
Holistic medicine, 251n
Hospitals, Islamic, 99–100, 103
Honorius of Autun, 165; Speculum ecclesiae, 270

ibn-al-Haitham. See Alhazen
ibn-Rushd. See Averroës
ibn-Sina. See Avicenna
India, contributions to mathematics, 117–21, 123–24
Indian Ocean, island world of, 34 (illus.)
Indonesian island world, 36 (illus.)
Industrial Revolution, 45
Infinity, arithmetical apprehension of, 120
Isaac the Jew, 103
Isidore of Seville, 165; Etymologies, 107
Islam, 129; cultural and scientific heritage, 93–99; hospitals, 99–100, 103; highly developed system of medical care, 99–104; pharmacies, 100; translations of Arabic texts, 104–11, 112–13; contributions to astronomy, 115; contributions to optics, 115–17; contribution to mathematics, 117–24

John VIII Paleologus (emperor of Byzantium), 31
John of Salisbury, 77, 83, 90–91; Metalogicon, 260–61
Julius II, Pope, 224
Jung, Carl Gustav, 139, 249
Justinian, Emperor, 114

Kant, Immanuel, 53, 249
Kepler, Johannes, 54, 116, 140, 141
Khayyám, Omar, 114

Michelozzo, 189; his lantern for cathedral of Florence, xix–xx (*illus.*), 1, 2n

Middle Ages: eclipse of science in early, 43; belief in transcendental universe, 56–58; culture achievements, 58–60; reawakening of science in later, 60–64, 132; philosophy, 78–79, 131; rationalism, 132–36; mysticism, 134, 138–41; contributions to astronomy, 142–45; medicine, 145–50, 154–56; observations of plant life, 150–53; alchemy, 177–87

Mona Lisa. See Vinci, Leonardo da

Montpellier, University of, 145

More, Thomas, *Utopia,* 20n

Moses, 85

Mukabala (al-jebr we'l), 127n

Müller, Johannes. *See* Regiomontanus

Mysticism, Medieval, 134, 138–41; revolt against, 144–45; in medicine, 145; symbols and multiple meanings in, 170; in Siena, Italy, 175–77. *See also* Alchemy

Natural-law philosophy, 52, 87–90, 226 and n

Nemorarius, Jordanus, 127

"Neolithic revolution," 45

Neo-Platonism, 226n

Newton, Sir Isaac, 81, 117, 124; and mysticism, 140; *Philosophiae Naturalis Principia Mathematica,* 141; and alchemy, 141–42

New World: speculation about, 20, 37; discovery of, 21, 23–24

Nicholas of Cusa, 2, 85, 86

Notre Dame, Cathedral of (Paris), gargoyle at, 166 (*illus.*)

Observatories, 114–15

Ocean River, Ocean Sea, 16–17, 22, 23, 33–39

O'Neill, Eugene, 27

Optics, Islam's contributions to, 115–17

Orbis terrarum, 16, 31

Oresme, Nicole, 199

Oxford University, 171n, 192n, 199; Bodleian Library of, 127

Padua, School of, 171

Paolo, Giovanni di, *Expulsion from Paradise,* 17 (*illus.*)

Paris, School (University) of, 113, 126, 192n, 199, 200n

Perspective, laws of, 210, 211n, 226–30

Petrarch, 80, 197; *The Renaissance Philosophy of Man,* 266–67

Peurbach, Georg, 2, 106

Pharmacies, Islamic, 100

Philip of Macedonia, 50

Philip of Tripoli, 111

Phlogiston, 186

Plant life, of Medieval observations, 150–53, 151 (*illus.*)

Platearius, Mathaeus, *Circa instans,* 152

Plato, 10, 17, 88n; his speculations about Atlantis, 20; Renaissance revival of, 31; translation of his dialogues, 33; and New World, 37; and Pythagoras, 51, 52; on harmony, 53–54; and Thierry of Chartres, 80, 81; *Timaeus*, 89, 107, 108n

Plethon, Georgios Gemistos, 30–33, 34, 40, 41

Pliny the Elder, *Natural History*, 107, 150

Pollaiuolo, Antonio, 12

Polo, Marco, 29, 37, 38, 40; his travels in the East, 27–28, 35; *Travels*, 28, 218, 257

Porrée, Gilbert de la, 81

Portugal, 2–4, early exploration of New World, 23–24, 27, 40–41

Poseidon, *See* Zeus of Artemision

Priscian, 74

Psychoanalysis, 249n. *See also* Jung

Ptolemy, 33, 50, 55, 70, 74, 77, 110; *Geography*, 22–23, 24, 34–35, 106, 144, 208n, 210–11, 211n; and map of Indian Ocean, 34 (*illus.*); on Southern Hemisphere, 38; *Almagest*, 106, 109, 112; *Optics*, 106, 116; and map projection, 212n

Pythagoras, 16, 70, 74, 89; philosophy of, 51–52; mathematics of, 52–54

Quintessence, concept of, 185 and n, 198

Raphael, *Disputa*, 11 (*illus.*)

Rashid, Harun al-, 100, 118, 120

Rationalism: Medieval, 132–36, 139; pragmatic, 140

Raziones seminales, 79, 82, 85

Razi (Rhazes), al-, 100–102, 103, 121; *al-Hawi*, 102; *Liber Almansoris*, 102, 109

Reconquista, 92–94

Regimen sanitatis Salernitanum, 147

Regiomontanus (Johannes Müller), 2, 106

Renaissance art, 192–98; conveying of physical space by, 198, 199–204; service performed for science, 204–10; exploration of third dimension, 204–14 *passim*; role in rise of modern geography and maps, 210–12; depiction of foreign travel, 218–19

Renaissance science, 191–204 *passim*; service performed by art for, 204–10. *See also* Science, modern; Scientific Revolution

Roger II (Norman King), 111

Roman Catholic Church, 76n

Roman civilization, breakdown in the West, 43, 50, 55, 58, 60

Romanesque churches, 158

Roman numerals, 121–22, 128

Rousseau, Jean Jacques, 249

Royal Society (England), 244n

Salerno, 145, 146–48, 156

San Lorenzo, Church of (Florence), 189 (*illus.*), 189–90

Sarton, George, 117. *See also* Bibliographical Index

Savonarola, Girolamo, 10–12, 11 (*illus.*)

Scarperia, Jacopo Angelo de', 22

Science, modern, 248–52; problem of controlling effects of, 242–43; prominent role of, 243–45, 248; growth of, 245–46; revival of, 247–48; link between birth of, and rise of capitalism, 248n. *See also* Renaissance science

Scientific Revolution, 80, 84–85,

Bibliographical Index

Jacobsen, T., *The Intellectual Adventure of Ancient Man* (with H. and H. A. Frankfort, Wilson, and Irwin), 257

Jones, H. L., ed. and trans., *The Geography of Strabo*, 257

Kantorowicz, Ernst, *Frederick the Second*, 263, 265

Keys, Thomas E., *The History of Surgical Anesthesia*, 263, 269

Kitto, H. D. F., *The Greeks*, 257

Klibansky, Raymond, "The School of Chartres," in *Twelfth Century Europe and the Foundations of Modern Society*, 260, 261

Koyré, Alexandre, *From the Closed World to the Infinite Universe*, 260–61

Kristeller, Paul Oskar: *The Classics and Renaissance Thought*, 262, 263; *Renaissance Thought II*, 263, 264, 265, 266; ed., *The Renaissance Philosophy of Man* (with Cassirer and Randall, eds.), 266–67

Latham, Ronald, ed. and trans., *Travels of Marco Polo*, 257

Lewis, Bernard, *The Arabs in History*, 262

McGarry, Daniel D., ed. and trans., John of Salisbury's *Metalogicon*, 260, 261

Mâle, Emile: *Notre-Dame de Chartres*, 259; *Religious Art From the 12th to the 18th Century*,

270; *The Gothic Image*, 270–71

Mallè, Luigi, ed., Alberti's *Della pittura*, 275

Markham, C. R., ed. and trans., *The Journal of Christopher Columbus*, 257

Marks, Robert W., ed., *The Growth of Mathematics*, 264

Métraux, Guy, ed., *The Evolution of Science* (with Crouzet, ed.), 260

Moody, Ernest A., "Laws of Motion in Medieval Physics," in *Toward Modern Science*, 258, 273

Moscati, Sabatino, *The Face of the Ancient Orient*, 257

Murray, Peter, *The Architecture of the Italian Renaissance*, 256, 274

Nachod, Hans, ed. and trans., "Francesco Petrarcho," in *The Renaissance Philosophy of Man*, 267

Obler, Paul C., ed., *The New Scientist* (with Estrin, ed.), 267

O'Malley, Charles D., ed., *Leonardo da Vinci on the Human Body* (with Saunders, ed.), 206n, 273, 275

Palter, Robert M., ed., *Toward Modern Science*, 142n, 257, 258, 262, 263, 268, 271, 273

Panofsky, Erwin, *Meaning in the Visual Arts*, 270, 271, 272–73

Paré, G., *La renaissance du XII*[e]

Picture Credits

Alinari/Editorial Photocolor Archives has supplied the illustrations on pp. xx, 8, 9, 14, 30, 32, 95, 157, 159, 163, 189, 200, 201, 203, 205, 207, 213, 215, 217, 218, 221, 223, 225, 227, 234, and 241. The fresco on p. viii is from the Anderson Collection of Alinari/Editorial Photocolor Archives.

Other illustrations are credited as follows: p. 3, courtesy Istituto Enciclopedico Italiano; pp. 6 and 11, Scala/Editorial Photocolor Archives; p. 17, Lehman Collection/Metropolitan Museum of Art; p. 19, courtesy Library of Congress; p. 34, courtesy British Museum, Map Collection; p. 36, map courtesy of Oxford University Press; p. 45, copyright © Paul Caponegro; pp. 46 and 47, Hirmer Photoarchiv, Munich; pp. 67 and 68, Helga Goldstein; pp. 72–73, from *La Cathédrale de Chartres*, by Etienne Houvet, 1926; p. 101, from *Illustrations miniatures*, Pierre Huard and Mirco Drazen Grmek, eds. (Paris: Les Editions Roger Dacosta, © 1960); p. 105, from *Geschichte der Medizin*, by Jean Starobinski (Lausanne: Editions Rencontre, and Erik Nitsche International, © 1960); p. 119, courtesy New York Public Library.

Illustrations on pp. 143, 149, and 151, are from *From Magic to Science*, by Charles Singer (New York: Dover Publications, 1958); p. 155, from *History of Surgical Anesthesia*, by Thomas E. Keys (New York: Dover Publications, 1963); p. 166, courtesy New York Public Library; p. 168, from *The Cathedral Builders*, by Jean Gimpel (New York: Grove Press); p. 172, Ing. Vincenzo Balocchi, Florence; p. 174, Grassi/Siena, in *Siena*, by Paolo Cesarini (Siena: Lombardi); pp. 179 and 181, courtesy University of Prague; p. 209, Syndics of Cambridge University Library.